Environmental Management and Decision Making for Business

Environmental Management and Decision Making for Business

Robert Staib

First published 2005 by
PALGRAVE MACMILLAN
Houndmills, Basingstoke, Hampshire RG21 6XS and
175 Fifth Avenue, New York, N. Y. 10010
Companies and representatives throughout the world

PALGRAVE MACMILLAN is the global academic imprint of the Palgrave
Macmillan division of St. Martin's Press, LLC and of Palgrave
Macmillan Ltd. Macmillan® is a registered trademark in the United
States, United Kingdom and other countries. Palgrave is a registered
trademark in the European Union and other countries.

ISBN 978-1-349-52072-5   ISBN 978-0-230-52446-0 (eBook)
DOI 10.1057/9780230524460

This book is printed on paper suitable for recycling and made from
fully managed and sustained forest sources.

A catalogue record for this book is available from the British Library.

Library of Congress Cataloging-in-Publication Data
Staib, Robert, 1994–
    Environmental management and decision making for business /
Robert Staib.
        p. cm.
    Includes bibliographical references and index.
        1. Industrial management–Environment aspects. I. Title.
HD30.255.S716 2005
658.4'03–dc22   2005043187

10   9   8   7   6   5   4   3   2   1
14   13   12   11   10   09   08   07   06   05

Transferred to Digital Printing 2011

Contents

List of Tables

List of Figures

Acknowledgements

I would like to acknowledge the support and help I have received over many years from people within various organisations. They have given me the opportunity to participate in and contribute to the development of better corporate environmental management practices and achievements. They have been from Rouse Hill Infrastructure Consortium Sydney Water Corporation, consulting engineers GHD Pty Ltd and the Sydney Olympic Coordination Authority in particular, and from the many consulting organisations, contractors, manufacturers and Government departments with whom I have worked.

I also acknowledge the use of unpublished material from these organisations upon which I have drawn to provide background information on current corporate environmental practices. Unpublished reports include the records of the audits and reviews I have conducted as an environmental manager (Staib 1993–2004; 1997–2000), the Sydney Water sustainability draft report (CSIRO, 2002), the Carlton United Brewery case study in Table 18.3 (Benn et al, 2004b) and the NSW Government economic appraisal guidelines (Aquatec Environmental Consultants, 1996). I would like to thank the International Association of Public Participation for allowing Fiona Court to use The Public Participation Spectrum (Figure 19.1). I would also like to thank the following organisations for permission to publish their diagrams: SAI Global for Figures 12.1 and 24.1 drawn from the International Organization for Standardization's environmental codes and the Global Reporting Initiative for Figure 14.3. The above material is listed in the reference section of this book.

In preparing this book I would like to thank the fifteen contributing authors for their help. They have given up their time freely and have brought a wealth of experience to the book. This experience is exemplified by the amount of material they have collectively published on the subject of environmental management and their continuing contributions to environmental education, training and management in Australia and overseas.

I owe a large debt to the staff and students from Macquarie University Graduate School of the Environment, Sydney Australia with whom I have been associated for twenty years. In conceiving this book I would like to acknowledge particular support from Peter Nelson, Alistair Gilmour, John Court, Richard Horsfield, Joy Monckton of Macquarie and Jacky Kippenberger and Stephen Rutt of Palgrave.

I would like to thank my wife Roslyn and children Natalie and Gregory for their patience, interest and feedback during the two years of the book's conception and development. It is our children and their children who will be the judges of our attempts to protect the world's natural environment while still providing a respectable quality of life for all people.

Contributing Authors

Carol Adams
Professor Carol Adams is Professor of Accounting and Head of School of Accounting, Economics and Finance at Deakin University, Victoria, Australia. She is a former Director and Council Member of the Institute of Social and Ethical AccountAbility and judge for the ACCA sustainability reporting awards. Her research is in social and environmental accounting and accountability.

Maria Atkinson
Ms Maria Atkinson is an Environmental Scientist and the Executive Director of the Green Building Council of Australia – a national not-for-profit industry association which promotes sustainable development and the transition of the property industry to implementing green building design practice and operations. Maria regularly presents at conferences and delivers public lectures on sustainable development locally and internationally.

Suzanne Benn
Dr Suzanne Benn is a biochemist and social scientist. She is a Senior Lecturer at University of Technology, Sydney, Australia where she researches organisational change for corporate sustainability and coordinates teaching programs in sustainable business.

Fiona Court
Ms Fiona Court is General Manager, Infrastructure Communications and Community Involvement for the NSW Roads & Traffic Authority. She is on the organising committee of IAP2 (International Association of Public Participation) Australasian Chapter. She has qualifications in both science and planning, and works to incorporate community initiatives into State Government planning.

Lorne Cummings
Dr Lorne Cummings is a Senior Lecturer in Accounting and Finance at Macquarie University, Sydney. He teaches advanced financial accounting theory and researches in international stakeholder theory, social and environmental accounting, and international financial reporting standards. He is a qualified Certified Practising Accountant and is a technical adviser on financial accounting to the Australasian Reporting Awards, the National University of Samoa and McGraw-Hill Australia.

Ken Cussen
Dr Ken Cussen teaches and researches in environmental philosophy and ethics at the Graduate School of the Environment, Macquarie University, Sydney, Australia.

Dexter Dunphy
Professor Dexter Dunphy is the Distinguished Professor, Faculty of Business, School of Management at the University of Technology, Sydney. Dexter has published extensively in the areas of generating sustainability and corporate change. He currently directs the Corporate Sustainability Project at UTS.

Andrew Griffiths
Dr Andrew Griffiths is a Senior Lecturer in the University of Queensland Business School, Brisbane, Australia. He has published four books on the topic of change and corporate environmental management. His work has been published in international journals, including The Academy of Management Review and the Journal of Management Studies.

James Guthrie
Professor James Guthrie is joint founding editor of the international research journal, *Accounting, Auditing and Accountability Journal,* and an editorial board member of 18 other research journals. He has published over 100 articles in both international and national refereed and professional journals, 30 chapters in books and has presented his research findings to over 200 national and international gatherings. Formerly Professor of Management at the Macquarie Graduate School of Management, he is now with Sydney University.

Craig McKenzie
Dr Craig Mackenzie is Head of Investor Responsibility with Insight Investment and was previously head of Governance and SRI at Friends, Ivory & Sime – now F&C Asset Management plc – where he led the creation of their Governance and SRI Team. He is a Senior Visiting Fellow in the Economics Department at University College, London.

Michael Polonsky
Professor Michael Jay Polonsky is the Melbourne Airport Chair in Marketing, within the School of Hospitality Tourism and Marketing at Victoria University in Melbourne. He has published widely on environmental and business issues, including editing several books on the topic.

Patricia Ryan
Emeritus Professor Patricia Ryan is an honorary professor at the Graduate School of Environment, Macquarie University, Sydney. She was previously

Head of Division and Professor of Business Law in the Macquarie Division of Economic and Financial Studies. She has taught and researched environmental law and management topics for over 33 years.

Robert Staib

Dr Robert Staib is an independent environmental management consultant with qualifications in mechanical engineering, business management and environment. He is a Visiting Fellow at the Macquarie University Graduate School of the Environment, Sydney, where he convenes courses in corporate environmental management.

Rory Sullivan

Dr Rory Sullivan is Director, Investor Responsibility with Insight Investment, London. He is the co-author of Effective Environmental Management: Principles and Case Studies (2001), the editor of Business and Human Rights: Dilemmas and Solutions (2003) and the co-editor of Putting Partnerships to Work (2004).

Ros Taplin

Dr Ros Taplin leads the Environmental Management Program within the Graduate School of the Environment, Macquarie University, Sydney. She has a research background in international and Australian environmental policy and politics and has a particular interest in climate change. Ros formerly was the Director of the Climatic Impacts Centre at Macquarie University and before that held positions at the University of Adelaide and RMIT University.

James White

Mr James White is Chief Analyst (Environment and Conservation Economics) at the Department of Environment and Conservation, New South Wales, Australia. He has qualifications in statistics, resource management and economics. His work involves regulatory impact analysis and environmental valuation. In 2001 he prepared New Zealand's national interest analysis for ratification of the Kyoto Protocol.

1
Introduction

Robert Staib

1.1 Outline

Great change is necessary in our society and in business organisations if we and the natural world are to live sustainably. We have started to change but it is slow and intermittent. Some sections of business are starting to embrace sustainable development with its three pillars: economic, social and environmental. I think we are in danger of losing the original emphasis on the *ecological* part of ecologically sustainable development by referring to the phenomenon as sustainable development. One of the justifications for sustainable development appears to be that we can sustain growth in sales and profit and not significantly impinge on the natural environment.

Sometimes this is hard to see when the population is heading towards 11 billion and developed countries are maintaining their materialistic bent and developing countries are striving to catch up. The change necessary is, one could say, unbelievably great but then again we must start somewhere. Large changes in society can take societies a generation or more to achieve as old ideas and people are replaced by new ideas and people. So we need to have at least a 20-year horizon and we need to educate our young people to new ways of thinking and new ways of seeing the natural world.

This book seeks to be part of the education of our future managers especially those managers of business organisations. This is an education so that they can learn to think strategically not only about future markets and sales but about the natural world and how much it means to humanity and how much their decisions can impact it for better or worse. I have been disappointed when I look at university web sites around the world at the peak business course – the Masters of Business Administration. I see how little the environmental subjects figure in the courses.

This book is an attempt to bring together in one place a range of information to help students of management to rethink their business approach, look 20 years into the future and try to envisage what the world can be like – if only we try. This book has arisen from two course units I convene at the Graduate School of the Environment at Macquarie University and my experience as an engineer, project manager and environmental manager in a variety of industries. In writing and editing the book I have been assisted by a number of very committed authors who think that sustainable effort is worth making.

1.2   Objectives

The primary objective for the book is as an environmental management textbook for tertiary students – in post graduate level environmental management courses though it should not be restricted to courses within environmental schools but should include other courses which I believe have a greater need for a well-rounded environmental management education: business, finance and engineering. In fact I am hoping it will be used by MBA students as well. It is written in a style that should make it attractive as a reference source for professionals and practitioners in the environmental management area. While it has mainly been written by Australian authors, we have attempted to make it suitable for an international audience by writing in a way that is universal with both theoretical and practical aspects covered.

The book is not a treatise on the subject nor is it the definitive document – who could write one anyhow? It makes use of some of the latest publications and thinking on the subjects that will hopefully give students an idea of the breadth of the subject. It exposes them to a range of ideas, which they can explore in more detail by consulting the many references cited. A short case study is provided at the end of many of the chapters. A list of questions is also included and this can be used as an assignment or to provoke further thought.

The contributing authors come from a variety of backgrounds in business and academia and each brings his or her unique perspective to the subject. In editing I have attempted to achieve some consistency of style without changing the authors' ideas.

1.3   Environmental decision making

I believe that a great number of environmental impacts are created (or embodied) when initial or early decisions are made, not when one gets involved in the later detail of a project or operation. Corporate decision making can be considered as a broad term that encompasses strategic planning and general management decisions and decisions made in the management of design – e.g. in the management of technical or engineering design of services and products. With the Sydney 2000 Olympics the biggest environmental impact was when Sydney won the games. The green Olympic strategy certainly reduced the potential impact but never eliminated it. One of the key aspects of management is decision making and environmental outcomes in business organisations are greatly influenced by corporate decision making. We have tried to keep this in mind in the book and expose the reader to many different approaches to decision making. We also discuss the organisational roles and responsibilities given and assumed by people and organisational teams, how individuals or teams can not only set environmental targets and facilitate their achievement but also along the way influence others to better environmental organisational practices.

1.4   Structure of the book

The book is divided into four parts: I. the environmental context of modern business; II. corporate processes and systems; III. culture and people; and IV. environmental management techniques and tools and then sub-divided into 25 chapters on specific topics. Figure 1.1 shows the framework that links the parts and chapters.

Part I of the book (Chapters 2 to 10) covers the *environmental context of modern business* by describing some of the external issues and influences that affect the environmental performance of an organisation. In Chapters 2 and 3 we discuss current trends in environmental impacts and in environmental attitudes and behaviour and in Chapter 4 environmental ethics and belief systems. In the next group (5 to 8) we look at some of the key government organisations and their processes which set the framework in which business organisations must work: politics and the environment; government environmental decisions; legislative and institutional frameworks; and environmental economics. We then broaden the discussion to include consumers and community (9). With the foregoing chapters as background we then discuss some of the ways in which the business community is responding to environmental issues and the issue of sustainability (10). I have refrained from defining business sustainability too closely but leave it to the reader to consider the term in light of the ways it is used by the different authors.

Part II of the book covers *corporate processes and systems* (i.e. internal organisational matters) starting with corporate environmental strategy (11) leading through environmental management systems (12), green marketing (13) financial management and accounting (14), corporate environment information (15) and finishing with social and environmental reporting (16).

Part III of the book covers organisational *culture and people* with discussions about organisational structures and roles (17), changing the environmental ethos of an organisation (18), and interaction with external stakeholders (19).

Part IV describes a selection of the *environmental management techniques and tools* that are available for organisations to use to help them become more environmentally responsible and achieve a high level of sustainability. It includes discussions of environmental decision making (20), environmental risk assessment (21), cleaner production (22), environmental design management (23), life cycle assessment (24), environmental impact assessment (25) and environmental auditing (26).

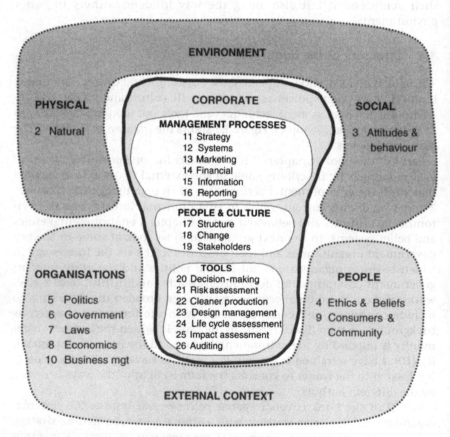

Figure 1.1 Environmental management and decision making framework

Part I

The Environmental Context of Modern Business

Part 1

2
Environmental Trends and Impacts
Robert Staib

2.1 Introduction

Environmental research is generating a large amount of information as we struggle to understand how natural systems work in their normal un-impacted state, how they respond to an ever increasing human presence and whether they will be able to absorb the effects of continuous harvesting and additions of pollution. Society has need of nature's resources to survive and to prosper though whether we need to use so much is being repeatedly questioned. Some authors (Hawken et al, 2000) believe the answer is to dematerialise society by doing more with less. Business organisations especially private organisations supply much of this material matter to our lives and in doing so use and utilise much of the natural resources of the world.

In this chapter we start with a selection of global indicators that show that the world's population is continuing to increase rapidly and that increased population is continuing to increase its average consumption per person of energy and material goods with significant global impacts on the natural environment. We also use two interesting indicators that measure the effect of humankind's ecological footprint on the planet: one shows how much we consume and the other how much is available – a demand and supply analogy. Unfortunately we are consuming more than is available and we are starting to whittle away our capital – our natural capital.

This living matter (or biosphere) has been the result of over 600 million years of evolution and the result is an intricate, though sometimes fragile, web of life covering our world – the animals and plants. The inanimate matter which has been forming since the birth of the world over 4 billion years ago is equally as important and supplies us with minerals, energy, water, soil and air although one could argue that the last three are living systems. The human race at about 1 million years (Noble and Davidson, 1997) is a neophyte in the world but is gradually usurping the others with its seemingly insatiable appetite for material things to satisfy both its basic physical needs and its psychological needs especially the elusive feeling of happiness (Hamilton, 2003).

Because business organisations are one of the main groups that supply matter to our lives, they need to understand these physical and psychological needs (or wants) and the consequences of continuing to supply materials in a profligate way. There is a lot to learn.

In this opening chapter we include a sample of this information and establish a global framework in which organisations can make strategic decisions about their market and its impact on the environment to significantly reduce the materialisation of society while still providing the products and services for a basic life and a meaningful human existence.

2.2 Trends and projections: population, energy, materials

We start our discussion with some of the trends that underlie or are driving environmental degradation. Increasing environmental impact can be thought of as a product of three major factors: *population* growth, *energy* use per person and *material* use per person (Ehrlich et al, 1977, p. 720).

Current world *population* (Figure 2.1) is about 6 billion, forecast to increase to about 9 billion in 2050 and stabilise at about 11 billion in 2150. The population in developed countries is growing a lot slower than developing countries and is expected to cease growing in about 20 years while

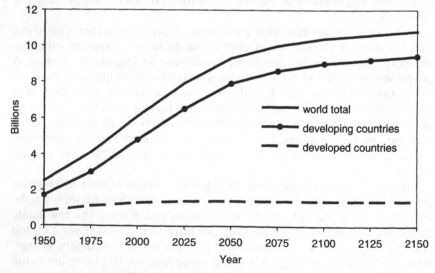

Figure 2.1 World population – projected
Source: United Nations (2002a), United Nations (2002b), Loh (2002).

the population in developing countries is expected to continue growing for a long time before it too is forecast to stabilise. A forward projection of 150 years is a bold step, but probably needs to be attempted.

Figure 2.2 shows the growth in world total *energy* production from 1950 to 2000 with a prediction to 2020. High-income countries use between 3 and 4 times the energy of middle and low-income countries combined (Loh, 2002). Continuing growth in use is forecast both in developed countries and in developing countries with major growth occurring in developing countries as they strive to bring their living standards to those of the richer countries (International Energy Agency, 2001).

We complete our trilogy of underlying trends with a discussion of *material* use with three indicators of consumption. There are many indicators but three will suffice: *food* consumption, *industrial* production and *water* use. Water is a bit problematic because it is a part of food and industrial production and probably represents a certain amount of duplication. Other missing aspects include agricultural and forest consumption for non-food products. One could also include consumption (by destruction) of the biosphere: native flora and fauna, soil and biodiversity.

Figure 2.3 shows increasing world *food* consumption. Consumption (per person) is forecast to increase by approximately 5% in developed countries but approximately 15% in developing countries by 2003 (United Nations, 2002a).

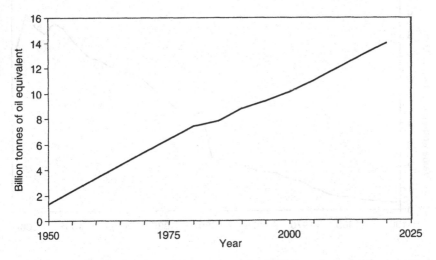

Figure 2.2 World total energy production – projected
Source: International Energy Agency (1998, 2001, 2003), Loh (2002).

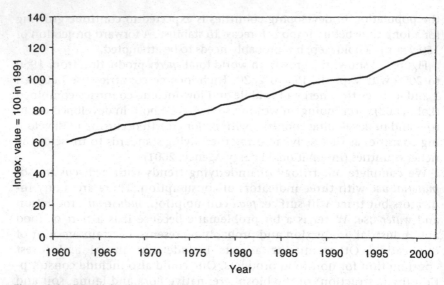

Figure 2.3   World food consumption
Source: World Resources Institute (2003). Index has 1995, 1996, 1997 trend smoothed.

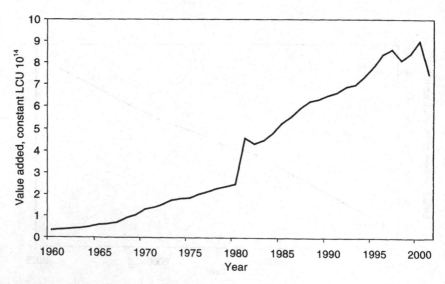

Figure 2.4   World industrial production
Source: World Resources Institute (2003); LCU = constant local currency.

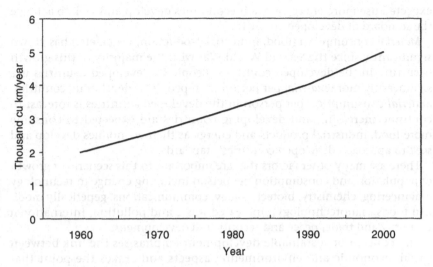

Figure 2.5 World freshwater withdrawals
Source: UNESCO (2003).

The world *industrial* production is shown in Figure 2.4 and a similar dichotomy exists between developed and developing countries as with the other indicators.

Figure 2.5 shows rising world *water* consumption. The developed countries use 2 times more water per person than developing countries. Developed countries use approximately the same amount per person for agriculture but significantly more for industry (Loh, 2002).

2.3 Drivers of increasing environmental impacts

The indicators (Section 2.2) of the trends in population, energy consumption per person and material consumption per person bring out some interesting facts.

The *population* of the world has increased about 3 times since the end of the Second World War (1945) with 90% of this growth occurring in the developing or lower income countries. It is forecast to continue growing from its present value of about 6 billion in 2000 and to stabilise at 11 billion in 2150 i.e. roughly 2 times its present size with most of this growth occurring in the developing or lower income countries.

Energy consumption has grown over 5 times since the Second World War with the majority of this growth occurring in the developed countries. Developed countries use about 3.5 times the *energy* per person than people in developing countries. *Energy* consumption per person in the developed countries is forecast to continue increasing and developing countries are expected use more energy as their economies develop and seek to achieve the standard of developed countries.

Material consumption (food, industrial production, water, etc.) has grown significantly since the Second World War with the majority of this growth occurring in the developed countries. People in developed countries use significantly more *material* per person than people in developing countries. *Material* consumption per person in the developed countries is forecast to continue increasing and developing countries are expected to consume more food, industrial products and energy as their economies develop and seek to approach developed countries' standards.

There are many other factors that are important to this scenario of growth of population and consumption per person including change in technology (engineering, chemistry, biotechnology, communications, genetically modified foods, nanotechnology), increased waste and pollution, international economy and trade, peace and security and development.

The concept of sustainable development emphasises the link between social, economic and environmental aspects and makes the point that environmental degradation cannot be stopped until the major issues of social equity and economic needs of all the world's people are met (United Nations, 2002a). This is acknowledged and is discussed in other chapters but since this book focuses on environmental aspects, we have not included these aspects in this opening chapter. The author wishes to place a strong emphasis on the environmental aspects because they may become diluted by the consideration of these other aspects in business organisations.

2.4 Some global environmental impacts

We have been selective in presenting some of the global environmental data with the objective of painting an easily understood framework. We have used the simple idea: increased environmental *impact* is a product of *population* growth, increased energy use per person and increased *material* use per person (Ehrlich et al, 1977, p. 720).

This idea simplifies very complex global interactions within countries and between people and the natural environment but maybe this is necessary. Each of the factors is growing (albeit differently in different countries and between developed and developing countries) but we can't escape the fact that the impact is more likely to be the product rather than the addition of the factors i.e. a 25% increase in each of the three factors could bring about an increase of almost 100% in the overall environmental impact.

We have not provided information on trends in environmental impacts (this is covered in detail in the literature) but Table 2.1 lists major global and regional environmental impacts in five categories and with some of

Table 2.1 Selected global environmental impacts and issues

Category or media	Major impacts	Some reasons for impacts	Cause or source: some examples
Air	Global warming & climate change	Green house gas emissions mainly carbon dioxide	Combustion, deforestation
	Human health risk from higher UV-B radiation	Ozone layer depletion	Refrigerants, solvents, fire fighting agents
	Ecosystem health risk	Pollutants, climate change, ozone depletion	See above
	Pollution	Emissions from transport, energy production, industry and agriculture	Combustion, waste disposal, evaporation of chemicals
Land	Pollution	Release of chemicals & fertilizers, waste	Industry emissions, farming practices, waste disposal
	Soil biological degradation	Clearing of vegetation, depletion of nutrients,	Land clearing, overgrazing, other agricultural practices
	Soil erosion	Clearing and destruction of vegetation, overgrazing,	Land clearing, urbanisation,
	Human health	Contaminated land	Intensive agriculture, industry, human use
Water	Depletion of resource	Over use, inefficient use, waste	Growing population, food production, industrial processes
	Pollution	Emission from industry, sewage disposal, agricultural practices, land clearing	Industry, agricultural fertilizers, soil erosion, chemicals, sewage
Natural & biological resources	Depletion of resource	Consumption of forest products, fish, animals, grasses, minerals, coal, oil	Land clearing, agricultural practices, pollution, industrial and domestic use
	Contamination or destruction of resource	Land clearing, agricultural practices, pollution, exotic species	Land clearing, agricultural practices, pollution,

Table 2.1   Selected global environmental impacts and issues – *continued*

Category or media	Major impacts	Some reasons for impacts	Cause or source: some examples
Biosphere	Destruction & loss of habitat, species, biodiversity	Changing land use	Land clearing, urbanisation, agricultural practices, pollution, waste disposal

Source: Gilmour (2004).

their reasons and causes. The listing is not meant to be comprehensive or to show a cause and effect relationship but is meant to illustrate the range of impacts and their interconnectedness. Many of the causes emanate from the actions of business organisations either by their operations and or by consumption of their products.

2.5   Global health: consumption and supply

Various indicators are being developed in an attempt to measure how human activity is affecting the health of the world's living and non-living matter. Figure 2.6 shows 2 indexes. Firstly ecological footprint measures humanity's use of renewable natural resources expressed in billion hectares

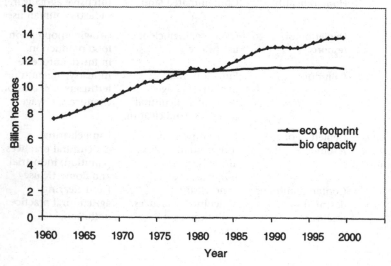

Figure 2.6   Global consumption and supply
Source: Loh (2002).

necessary to produce food and fibre, sustain energy and material consumption and give space for infrastructure i.e. it is a measure of *consumption* or demand. Secondly bio-capacity measures the ability of the planet to produce these renewable natural resources i.e. it is a measure of *supply*. The graph shows that in the early 1980s *consumption* (eco footprint) started to exceed renewable *supply* (bio capacity) and we started eating into our capital – in this case the planet's natural capital (Loh, 2002).

A third index, the living planet index is also used to measure the state of natural systems and is the average of three sub-indices measuring changes in forest, freshwater and marine ecosystems and its trend is similar to the widening gap between the eco-footprint and bio-capacity of Figure 2.6 (Loh, 2002).

2.6 Future generations

If one looks backward 4 or 5 generations to the early 1900s one is confronted by a vastly different world in terms of population, energy and material consumption. Who could have, or did, predict the changes that have occurred in the 20th century? With this history in mind one could look forward and question the reliability of the environmental predictions of the next 100 years. Two groundbreaking modelling studies of the environmental world of the future (Meadows et al, 1992, 1972), looked about 100 years into the future to 2100 – about 4 or 5 generations. Many of the forecasts made in the first study of continuing environmental degradation and overuse of the earth's resources are still valid (Meadows et al, 1992 p. xci). Despite their prediction of the potential for a continuing environmental decline or a global environmental catastrophe they believe that a sustainable society is still technically and economically possible.

2.7 Summary: implications for business organisations

In the chapter we have presented global environmental data showing trends in the growth of the world's population and the growth in consumption per person of energy and materials (food, industrial products and water). We have listed many of the types of environmental impacts that have and will result from the growth in these factors and finally we have presented global data showing how this is impacting the planet. The data is telling us that we have reached and are exceeding the productive capacity of the planet's living resources and that we are diminishing its stock of living resources – the planet's natural capital.

We have used the simple idea that increased *environmental impact* is the product of *population growth*, *increased energy* use per person and *increased material* use per person.

Most business organisations will have little control over population increase but they can control the energy and material they embody in the production of their products and services and the energy and material used in the use and disposal of their products. Therefore the challenge to business organisations is to dematerialise society to save and preserve our living planet (Hawken et al, 2000).

2.8   Questions

- Discuss and review the following statements. Many of the predictions of environmental catastrophe are open-ended and do not take into account feedback loops e.g. humanity when faced with diversity has always reacted intellectually and responded with solutions. There are signs that a lot of the adverse environmental trends are improving and the world is developing institutions and arrangements to better manage global environmental issues e.g. United Nations Environmental Program, World Business Council, Kyoto Protocol on climate change. Refer to Bailey (2000) and (United Nations, 2002a).

- Should we be responsible for future generations (1, 2 or more generations) or should we be responsible for our own generation and let future generations solve their own problems as they arise bearing in mind that our predictions of the future may be completely wrong?

3
Environmental Attitudes & Behaviour

Robert Staib

3.1 Introduction

This chapter discusses how public environmental concern has fluctuated since the Second World War – sometimes increasing, sometimes decreasing and reviews some of the literature on environmental attitudes and behaviour.

Chapter 2 documented some significant environmental impacts and trends. The real factors that underlie these trends may not all be known to the public but this does not stop public concern being raised about environmental problems or issues. An understanding and knowledge of how and why public environmental concern changes is important for business organisations. Firstly increased public concern about environmental issues can lead to governments passing new legislation and changing regulations and policies (Staib, 1998). Secondly it can lead to change in people's buying habits (Chapter 9). These effects can have important follow-on effects for business organisations especially for their medium term strategies.

Environmental concern is commonly measured through the use of opinion polls that sample people's views e.g. posing a question like "What is the most important issue facing society today?" The opinion poll questionnaire may present alternatives like unemployment, economy, education, health, crime, environment or respondents may nominate issues. The analysis may show that the concern for the environment is ranked third overall or that 20% of the respondents say that environment is an important issue. Similar questions asked over many years can show changes and trends. For the graphs in this chapter I have not used the data that shows the ranking of environmental concern relative to other concerns but have used the level of concern for environmental issues. A selection of opinion poll results mainly from the USA and Australia is used to illustrate the issues.

Concern: attitudes and behaviour

The term *concern* is used generically in this chapter but we discuss whether the *concern* is a measurement of *attitude* or *behaviour*. People may say that

17

they are concerned about the environment (have an *attitude* towards or about) but will they take action (change their *behaviour*) to make fewer environmental impacts? A common assumption is that a change in *attitude* inevitably results in a change in *behaviour*. This is discussed with research showing that establishing a direct link between the two is often difficult and sometimes tenuous. Latest research proposes more elaborate theoretical and measurement models that include another process called *behavioural intention* in an attempt to allow better prediction of likely behavioural change.

Finally the implications of the opinion poll values and trends and the attitude behaviour research for business organisations are discussed.

3.2   Concern for environmental issues – public opinion polls

In this chapter I use the example of pollution because it was one of the early manifestations of environmental concern. From the late 1960s to the early 1990s in the period of rapid economic growth after the Second World War, a combination of increasing world population, energy use and industrialisation have contributed to the production of a greater quantity of pollutants discharged into the environment (Figures 2.1, 2.2, 2.4 and Table 2.1).

These pollution problems have been significant for industrialised countries e.g. United States of America (Milbraith, 1985) Japan (Miyamoto, 1991), Great Britain (Clapp, 1994) and Australia (Coward, 1988).

This has resulted in a worldwide increase in the concern for the environment and the effects of pollution since the late 1950s and evidence of this increase is discussed and illustrated with data from opinion polls and other indicators in Australia and other industrialised countries. Firstly in this section we present some data from United Nations, USA and Canada and then in Section 3.3 some Australian data.

Soroos (1981) studied trends in the debates on environmental topics in the United Nations and showed an increase in number from 1968 to 1972 and a decrease to 1976 but not to the level of 1968. Dunlap (1991a) analysed 23 separate public opinion surveys undertaken from 1965 to 1990 in the USA on attitudes towards the environment. No survey extended continuously over the full period. Each poll extended over a number of years allowing some assessment of trends. The questions asked by each poll varied but each attempted to measure the level of public concern for environmental issues.

Figure 3.1 plots the average values of the 23 surveys described by Dunlap (1991a). These averages should be interpreted with caution especially in the early years when there were only a small number of polls. The vertical axis shows the percentage of respondents who expressed concern for the environment. The average of the polls shows a rise to 50% in the late 1960s, a lower concern during the 1970s (around 30%) and a gradual increase to about 55% by 1990.

Dunlap (1991a) refers to the 'issue attention cycle' model (Downs, 1972) in assessing the trends in public opinion surveys shown in Figure 3.1. Downs was one of the first authors to identify a cycle involved in the solution of large scale public problems in particular environmental problems. His article suggests that key domestic problems in the USA often leap into prominence, remain for a short time and then generally faded from the public attention largely unresolved. Dunlap says that the decline in interest in the environment predicted by the Downs model was not strongly evident in the surveys and that support for the environment remained high. He says that although there was strong public concern for the environment this did not automatically translate into the social changes needed for solving major environmental problems i.e. the link between change in concern (attitudes) did not directly correlate with social (behavioural) change.

Other data is consistent with these USA trends. McGeachy (1989) studied trends in magazine coverage in the USA over the period 1961 to 1986 and showed a sharp build up in number of articles on environmental issues to 1970 and then a decline though not to the level before 1970. Holdgate et al (1982) used the number of articles in the New York Times on environmental topics in the years 1960, 1970, and 1979 to show a similar pattern. Dunlap (1991b) in a further study of trends in views on environmental issues, showed significant increase in concern for the environmental from

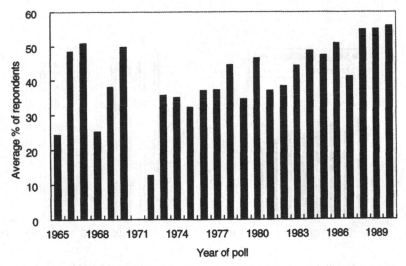

Figure 3.1   USA opinion polls: concern for the environment – % of respondents
Source: Dunlap (1991a). Average of 23 polls.

early 1982 to 1990. Parlour and Schatzow (1978) documented a significant build up in newspaper coverage of environmental and pollution issues from 1960 to 1970 in Canada.

3.3   Environmental issues – Australian opinion polls

Australian opinion polls measuring environmental concern from 1975 to 1994 and then 1990 to 2002 are reviewed below. There was no continuous poll over the period and therefore no consistent measure over the full period (Lothian, 1994, 2002). Figure 3.2 shows a downward trend (Morgan Gallop) in concern for the environment from 1975 to the mid 1980s and a high level in the Newspoll from early 1990s to 2002. Although these polls are different, Lothian (1994) believed that the level of public concern during the 1980s was relatively stable between the two higher periods of higher concern in 1975 and 1990. This is similar to Canada and USA data from Section 3.2. Lothian (1994) believed that although these opinion polls showed strong concern for the environment, governments seemed to be held back because of a perception that there was insufficient support. This is a similar conclusion to Dunlap (1991a).

In contrast to the Newspoll's consistent 60% of Figure 3.2, polls of the Environment Protection Authority of New South Wales Australia (EPA, 2000a; EPA, 2003) show a downward trend in concern for the environment relative to other social issues for the period from 1994 to 2003.

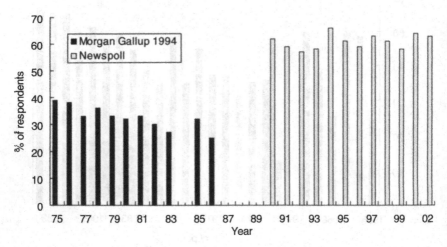

Figure 3.2   Australian opinion polls on concern for the environment
Source: Lothian (1994, 2002).

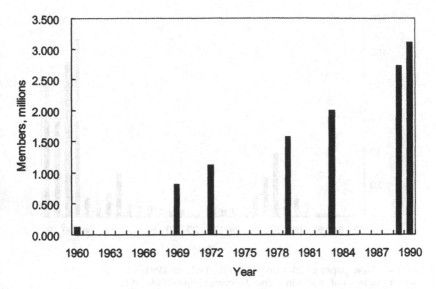

Figure 3.3 USA environmental group membership
Source: Mitchell et al (1991).

3.4 Growth of environmental lobby groups

Another indicator of changing public concern for the environment is the rise in the number of members of environmental interest groups. Figure 3.3 shows the steady build up of members in national environmental lobbying organisations in the USA (Mitchell et al, 1991) from 1960 to 1990 despite the up and down trends of the public opinion polls in Figures 3.1. There is probably insufficient data to be sure that membership was continuously rising in this period. There was also a steady build up of membership in three of the largest environmental interest groups in NSW Australia from 1960 to 1978 (Craney, 1980).

3.5 Concern for environmental issues – newspapers

In this section we discuss the use of newspaper coverage of environmental issues as a measure of environmental concern with the objective of further illustrating how concern fluctuates over time. We use Australian data though as mentioned earlier similar data exists for other industrial countries.

Figure 3.4 shows the number of articles on pollution (air, water, other) in the Sydney Morning Herald (a major daily morning newspaper in Sydney Australia) (SMH) from 1963 to 1991 (Sydney Morning Herald, 1964 to 1991). The number of articles on a particular topic is assumed to reflect the

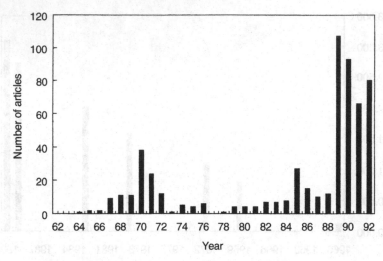

Figure 3.4   Newspaper articles on pollution, 1962 to 1991
Source: Manual index of articles in Sydney Morning Herald (1964–1991).

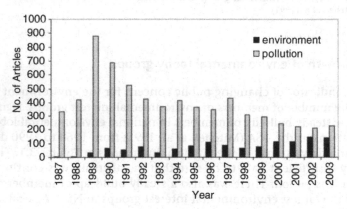

Figure 3.5   Newspaper articles on environment and pollution, 1987 to 2003
Source: Computerised index of Sydney Morning Herald (2004).

level of public concern as perceived by the newspaper journalists and editors. It shows a build up of the number of articles in the late 1960s peaking around 1970, a reduction in the number of articles during the 1980s though not down to the pre 1970 level. Towards the close of the 1980s there was a build up of the number of articles to a peak in the early 1990s which is the extent of this first set of data. The peaks around 1970 and 1990 roughly corresponds to the peaks in Figures 3.1 and 3.2. Hoffman (2001, p. 144) produced a similar graph for the USA with peaks around 1970 and 1990.

Figure 3.4 is based on a manual index of articles in the SMH. Figure 3.5 extends the data based on a computerised database of the articles from 1986 to the 2003 (Sydney Morning Herald, 2004). The data in Figure 3.5 'no. of articles' is the number of SMH articles which include the word 'pollution'. It is not as focused as the manual index, which was probably selective in identifying relevant pollution articles. The data in Figure 3.5 is probably overstated and could include uses of 'pollution' other than associated with the environment. This may explain why the two data sets are not consistent. Figure 3.5 is included though to show the trend, which is a reduction from the early 1990s to the present. This is consistent with the downward trends identified by the NSW EPA (EPA, 2000a; EPA, 2003).

Notwithstanding the trend of 'pollution' in Figure 3.5, there is still much written in the SMH about environmental and sustainability issues particlarly with an emphasis on the use of energy and resources. Figure 3.5 histogram for 'environment' shows the number of articles on general environmental issues. This data is the occurrence against the sort criteria: '(environmental or ecolog*) and sustainab* not pollution' which excludes the 'pollution' data and most other meanings of environment e.g. legal environment. It suggests that concern from pollution may be being replaced by broader environmental concerns e.g. environmental sustainability.

3.6 Concern for the environment – summary

The data presented in this chapter suggests that the trends in concern for (or attitudes about) the environment and pollution in Australia are similar to those in other parts of the world. The broad picture is that the late 1960s and early 1970s were characterised by rising concern for the state of the environment and through the 1970s and 1980s there was a falling and then levelling of concern. During the late 1980s and early 1990s there was a significant rise in concern. It is suggested that the rise in concern in the early 1970s was a concern for regional pollution brought about by a large rise in industrial activity, use of energy and population after the Second World War. A large part of the rise in concern in the late 1980s and early 1990s seems to have been caused by the increasing impact of global pollution as distinct from the earlier concern for regional pollution (Staib, 1997, p. 22).

The data in Figure 3.5 suggests that as pollution has been brought under control in western industrial nations there is less public concern for it. In contrast there appears to be a rising concern for the broader issues of sustainability as population and energy and resource use continue to grow. An understanding of these trends by business organisations is important as input into the strategic planning of business operations.

Hoffman (2001, pp. 144–146) in his book 'From Heresy to Dogma; An Institutional History of Corporate Environmentalism' studied the impact of environmental issues on industry. He believed that the corporate attention to the environment closely matched public opinion and that industry attention tended to follow the ups and downs of public opinion rather than the upward trend of environmental regulation and expenditure.

3.7 Public opinion and public policy

The previous sections discussed trends in concern for the environment and several authors alluded to the fact that this increase in concern did not necessarily translate into governmental action on the environment. In the next few sections we explore the link between change in *concern* and resultant *action* using the terms *attitude* and *behaviour* favoured by the psychological literature and discuss some ideas from this literature. Relevant questions are therefore: do changed attitudes towards the environment bring about changes in behaviour that can reduce harm to the environment; can a measurement of changed attitudes be used to predict changes in behaviour; and if one changes people's attitudes will their behaviour towards the environment change?

Luttbeg (1968), in reviewing models that attempted to explain how public opinion (attitude) is converted into public policy (behaviour) in the American context, described many influential factors including: individual attitudes, opinions, actions; public attitudes, opinions, actions; method of determining or perceiving public opinion; leaders perception of public opinion; individuals perception of leaders performance in reflecting or acting upon public opinion; role of the political party as a reflection of individual or public opinion; role of pressure groups and their relation to the individual and to the leaders; role of powerful individuals; role of other society groups; and role of politicians as trustees of their constituents.

The strength of public opinion can determine political agendas and the pace of policy implementation (Rosenbaum, 1991). 'With heightened concern (for the environment) politicians and business leaders are likely to give environmental considerations greater emphasis in their decision-making' (Parris, 2003). Staib (1997, 1998) showed that increased concern about an environmental issue by the public (measured by number of articles on the environment in a metropolitan newspaper) often comes before increased political action (measured by numbers of pages of recorded debates on the environment in Parliament).

Rajecki (1990) suggests that attitude change can be brought about by: direct experience; a persuasive message in the media; communications with acquaintances; or various forms of education. It may be thought that if attitudes change a corresponding behavioural change will follow but this is not always so.

3.8 Attitude and behaviour models

By the late 1960s research findings were not fulfilling earlier prophesies of a strong nexus between *attitudes* and *behaviour*. In 1970s researchers started to look more closely at what had been measured in the earlier studies. Many studies have shown in certain situations that a direct link between attitudes and behaviour is sometimes very hard to establish e.g. environment (Rajecki, 1990), environment and marketing (Hini and Gendall, 1995) and transport (Golob and Hensher, 1998).

Behavioural intention

Other studies discussed below have attempted to produce better models that can be used to make predictions about behaviour. An additional process *behavioural intention* was introduced and found to correlate better with measured behaviour particularly a single behaviour (Ajzen and Fishbein, 1977, 1980; Ajzen, 1982).

Global attitudes and global behaviours

Weigel and Newman (1976) initially measured people's attitudes to environmental and pollution issues, then using field situations they measured the same people's behaviour through approaches such as whether people would: sign a petition demanding action of legislators; participate in a litter pick up themselves; and recycle their own waste. The average correlation between a global index of measured attitudes and individual behaviours was 0.32, with categories of behaviour (petitioning, litter pick up, and recycling) the correlation was 0.42, and with a general index of behaviour was 0.62. This illustrated that global measures (or multi act measures) of attitudes and global measures of behaviour were more likely to show a positive correlation.

Correspondence entities

Ajzen (reported in Rajecki 1990) showed that there needs to be a correspondence between the behavioural entity measured and the attitude entity measured for a significant correlation to exist. His team defined four entities for both attitudes and behaviours: *target* – the attitudinal object, *action* – what one would like to do with the object; *time* – in which both exist; *context* – the situational reference. They postulated that, for maximum correlation between the measurement of attitude and the measurement of behaviour, the correspondence between each of the entities needs to be close. In a study of 109 published investigations within which a total of 142 attitude behaviour relations were reported they found support for the correspondence theory.

Behavioural intentions, cognitions and affective responses

Zimbardo and Lieppe (1991) proposed a more detailed model to help describe the attitude behaviour nexus by introducing further concepts like

BEHAVIOURAL INTENTION I would give a beer to a responsible 18 year old. I intend to vote for a referendum to lower the drinking age.		**BEHAVIOUR** I argued for an 18 year old drinking age in a conversation with my friends. I wrote to my local member protesting about the drinking age.
	ATTITUDE I am in favour of reducing the legal drinking age.	
COGNITIONS Lowering the drinking age will reduce the consumption of hard drugs. Alcohol in moderation is an important social behaviour.		**AFFECTIVE RESPONSES** I enjoy drinking socially. It angers me that 18 year olds can be drafted but cannot drink socially.

Figure 3.6 Example of an attitude behaviour model – under age drinking
Source: Zimbardo and Lieppe (1991); Ajzen (2001) – when the drinking age was 21.

behavioural intentions, cognitions and *affective responses* (Figure 3.6 is an example). He believed that attitudes can guide behaviour and measurement of attitudes can be used to predict behaviour, but careful attention is required to what attitude is being measured and what specific behaviour is being predicted.

Attitudes, behavioural intentions, behaviour

Golob and Hensher (1998) in studying attitudes towards greenhouse gas emissions and transport behaviour moved away from a linear model and used feedback paths between each of the three processes (*attitudes, behavioural intentions, behaviour*) and assessed the relative strengths of each path (Figure 3.7). They found good correlations between the attitudes and behavioural intention and between behavioural intention and behaviour. They also described the concept, not insignificant, where subsequent behaviour can also influence or change previous attitudes.

Moving on from the issue of general public attitudes we briefly touch on the subject of advertising. Business organisations rely on their ability to persuade people through advertising to buy their products. Rajecki (1990)

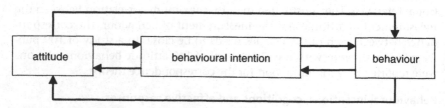

Figure 3.7 Attitude behaviour model
Source: Golob and Hensher (1998); Ajzen and Fishbein (1977).

says that there is considerable research on the relationship between attitude change and behavioural change in this area. He says that change is measurable but the changes in behaviour are minor shifts not large shifts. Notwithstanding this, he says the minor shifts can be significant for a person or community.

3.9 Direct experience

Before we summarise the attitude behaviour issues it is worth looking at the concept of direct versus indirect experience. Fazio and Zanna (1981) states that direct experience is more likely to develop stronger attitudes than indirect experience. The rationale is: 'direct experience may simply make more information available and hence result in a more accurate and stable attitude; direct experience may cause the people to focus on their behaviour during that experience and thus this behaviour may become an element on which to base an attitude; direct experience of a voluntary sort may involve repetition or rehearsal and may lead to an attitude that is more easily retrieved from memory.'

Zimbardo and Lieppe (1991) said that concern for the environment can be a matter of 'out of sight, out of mind.' People are more concerned about pollution when they can see it or it affects them. When the symptoms of the problem disappear the concern can lessen. It may also be that when people perceive that somebody (e.g. government or industry) is pursuing a strategy to alleviate the environmental problem, there is no longer a need to be concerned.

Staib (1997, 1998), in developing a process model to explain how major regional pollution problems have been solved, demonstrated the importance of direct experience by showing that people are more likely to become concerned about environmental pollution if it affects them personally. When enough people become affected and show their concern to the government, governments are more likely to act.

3.10 Attitude behaviour summary

The discussion in previous sections suggests that there is a link between attitudes and behaviour though not a direct one. To use or measure attitudes as a predictor of behaviour one needs to carefully consider the current research some of which is cited in this chapter and summarised in Table 3.1. It is necessary to understand and use the more detailed attitude-behavioural models as shown in Figures 3.6 and 3.7. Organisations need to be aware of the considerable amount of social psychological research that is available to address the issues of attitudes and behaviour.

Table 3.1   **Social psychological principles of the attitude behaviour nexus**

Principle	References
Single measures of attitudes should be used to measure single behaviours.	Weigel and Newman 1976
Multiple act measures (global) of attitudes should be used to measure multiple act behaviours.	Weigel and Newman 1976
There should be a direct correspondence between measures of attitude and behaviour for each of the entities: action, target, time, and context.	Ajzen and Fishbein 1977
Direct experience of an object is more likely to develop stronger attitudes towards that object and as a consequence lead to stronger behaviour towards that object.	Fazio and Zanna 1981 Zimbardo and Lieppe 1991 Staib 1997
An attitude-behaviour model should consider behavioural intent, cognitive and affective response	Ajzen and Fishbein 1977 Zimbardo and Lieppe 1991 Ajzen 2001

3.11   Implications for business planning

In this chapter we have discussed different ways that have been, and can be, used to measure public environmental opinion (*concern for the environment*). We have discussed how public opinion can change over time and can influence political action and the response needed from business organisations. Political action (e.g. new legislation) and public opinion can directly and indirectly affect an organisation's operations especially in the medium term. In a related topic, Chapter 9 discusses how the behaviour of consumers can affect an organisation's market.

We have also discussed the relationship between *concern for the environment* and environmental *actions* using the psychological terms *attitude* and *behaviour* to access the considerable body of psychological literature on the subject. We have illustrated that the link between *attitude* and *behaviour,* while important, may not be as direct as one might suppose. So for an organisation to use this psychological research and the information from opinion polls, it needs to understand some of the principles we have discussed in this chapter.

For effective medium term and strategic planning it is important for business organisations to understand the way changing public concern for the environment (*attitudes*) and changing public actions (*behaviour*) may ultimately affect their operations. They need to plan accordingly for as Hoffman's book title (2001) suggests, 'yesterday's heresy can become tomorrow's dogma'.

Case Study: direct experience & stronger attitudes

Social psychologists (Fazio and Zanna, 1981) found that direct experience is more likely to develop stronger attitudes than indirect experience e.g. people are more likely to react to pollution when it has affected, or is about to, affect them personally especially where the effects are visible. This is illustrated by the history of the treatment and disposal of sewage in Sydney, Australia. As the population of Sydney expanded from 300,000 in the 1880s to over 4,000,000 in the late 1990s, the problems of disposal of sewage (a major contributor to water pollution) increased.

In general terms the sewage was disposed of at locations that were away from population centres e.g. to rivers, then to the harbour, then to the land, then close to the ocean cliffs and finally to deep ocean outfalls (Coward, 1988; Staib, 1997). Coward's appropriately titled book *Out of sight, Out of mind* illustrates a recurring pattern in the identification and the resolution of Sydney's water pollution problems over 150 years of European settlement. As the population of Sydney expanded, people were more likely to come into contact with the effects of sewage disposal (through the location of their dwellings and their places of recreation or work) and when the pollution affected people personally they were more likely to seek political action.

Staib (1997, 1998) showed that public concern increased, and subsequent political action occurred, as more and more people became personally affected by the sewage pollution problem. The public concern started to increase from the time of first identification of the pollution problem until some political action was taken.

3.12   Questions

- Find some recent surveys of attitudes about (or concern for) the environment and suggest how these attitudes and resulting behavioral change could affect the products or services your organisation provides.
- Design a survey that measures attitudes, behavioural intent and behaviour and describe a method to establish the correlation between them – suggested reading Golob and Hensher (1998) and Hini and Gendall (1995).

4
Environmental Ethics and Belief Systems

Ken Cussen

4.1 Introduction

This chapter discusses the environmental relationships between individuals, business organisations (or corporations) and nature. It does this by
outlining belief systems based on distinctions between: indigenous and
western cultures; and anthropocentrism and ecocentrism. The chapter
closes by linking these belief systems to sustainability.

4.2 The corporation and society

The form of social organisation we call a corporation had its origins in the
great medieval trading cities of the Mediterranean and northern Europe
and developed into its modern form concurrently with the development of
the industrial revolution. From their beginnings, corporations were not
free-standing entities divorced from social needs and demands but rather
were immersed within, dependent upon, and responded to complex and
interlocking environments: the market, the social (ethical/political/legal)
and the natural. All three provide opportunities for wealth creation and
place a variety of constraints on corporate behaviour. Thus it has been
recognised since the inception of corporations that they have wider responsibilities than simply to return a profit to shareholders. The precise
nature of those responsibilities has been in a state of flux and it is only relatively recently that the natural environment has begun to gain a similar
level of attention to the market and the social environments.

In the last few decades, since the release of the Brundtland Report
(Brundtland, 1987), and development of ideas like Ecologically Sustainable
Development and Corporate Social Responsibility (or Triple-Bottom-Line
accounting) that these environmental constraints have become more
demanding and explicit and the need for corporations to respond to them
more urgent. It has become widely recognised that corporations, through
improved management structures and techniques, can and should fulfill an

important function in lessening their environmental impacts. There are reasons why they should do so: internally because employees and shareholders have an expectation that their corporation will act in an environmentally responsible way; and externally because of pressures from customers, environmental activist groups, ethical investment funds, concerned individuals and environmental laws.

While this is increasingly well-recognised in modern corporate culture, there remains a problem. As with all moral questions, the most important reason for a corporation to behave in an environmentally responsible manner is simply that it is the right thing to do. No amount of internal or external pressure can substitute for the determination to behave ethically – but there has emerged no clear ethical consensus about the values and reasons underlying environmental concern.

In a world dominated by consumerism, the notion of economic growth and the rationality of the market, it is easy for environmental interests to be disregarded, or to be seen as a limiting factor on the freedom of corporations and individuals to make a profit. This is partly because there has been an imbalance in the way that *values* have been conceived and attributed to the market on one hand and the environment on the other. The values underlying the market are often clear and beyond dispute. They can be articulated in terms of the value of wealth creation, jobs, profit, export income and the right of individuals to prosper from the exploitation of their property. All these are things of real value that undeniably contribute to human well-being and cannot lightly be dismissed.

The problem lies in the fact that the values underlying environmental concern, being of relatively recent origin, have been less clear. Against the background of well-established values in the market, advocates for the environment have often been forced onto the defensive. A vital step towards giving the environment a voice, is to be able to articulate a concise, logically compelling and plausible case for the preservation of the natural environment, a case that is capable, at least in principle, of being given equal weight with the values underlying market activities.

This raises the question of how corporations *ought* to respond to this new level of environmental demand. How is a corporation to determine and build into its decision making, its ethical responsibilities for and duties to the natural environment in a situation of market, social and ethical flux? While no simple answer can be given, it is important that people making decisions affecting environmental outcomes understand the range of environmental ethical positions in the community and the conceptual frameworks that determine the human evaluation of environmental matters.

To understand the underlying values and conceptual ethical frameworks involved in environmental concern the corporation needs to understand some of the historical and conceptual background to our present situation.

4.3 Humans, nature and culture

An objective observer might reasonably expect to find that all humans on the planet, having common evolutionary origins, survival needs and springs of happiness, would also have a common attitude to, and place a common value on, the biological system from which they sprang and which supports them. This observer would not of course have any *a priori* means of knowing what that attitude would be: for all he would know before investigating the matter, humans might revere nature, be contemptuous of it, or be indifferent to it. But he would reasonably expect to find a common attitude because humans share a common biological relationship to the natural world.

He would be wrong. He would soon notice that overlaying the biological facts are the forces of diverse human cultures. One of the chief things that marks the differences between cultures is precisely their conception of their relationship to, and their treatment of, nature (Cussen, 2002).

4.4 Indigenous cultures

Visiting indigenous groups in northern Australia or the Americas, he would discover there groups of people who do not have the concept of 'nature' conceived as an entity distinct from themselves; rather they think of themselves as part of a unified whole, a view that entails a deeply reverential attitude to their biological and geographical surroundings (Knudtson and Suzuki, 1992; Rose, 1996). Their social structures, kinship arrangements and meaning of life are, for them, so intimately tied to their natural surroundings that to be made an outcast from the group means that the sacred elements of nature have also withdrawn their favour and protection. In these societies the question of how one *ought* to treat nature does not arise: the individual's duties and obligations both to other people and to the natural world are determined by tradition, by his/her place in the kinship web and were probably decided thousands of years before the individual was born. This relation to nature is repeated in various societies around the planet.

4.5 Western cultures

This relation is not universal. Our observer would soon discover at the other extreme cultures whose behavior indicated a very different conception of their relation to nature. He would find a vigorous culture that broadly identifies itself as *western*, or modern or industrialised and defines itself against those cultures that it characterises as *traditional* or less civilised. Unlike indigenous cultures, western cultures make a distinction between the human and natural parts of the world and that distinction, and the conceptual distance it places between themselves and the natural world, has allowed them to use nature to increase enormously their productivity, wealth and population and to dominate and marginalise these non-western cultures.

Our observer would soon discover that, as well as differences between how cultures conceive of, relate to and place value on the natural world, there is also a curious cleavage of opinion and behaviour *within* the western cultures. While most western humans, when questioned, profess to having a certain fondness for nature, their behaviour reveals deep differences in their evaluation of nature. Some people in the industrialised cultures care very deeply that nature should be preserved in as natural a state as possible while others would prefer to exploit economic benefit from it to enhance human well-being, even at the expense of degrading its natural character. These value conflicts are reflected in the spectrum of moral theories that have been developed for regulating the human relationship with nature (Norton, 1992).

4.6 Anthropocentrism

Western ethical conceptions regarding nature derive largely from two ancient sources: the Judeo-Christian tradition and Greek philosophical reflection, especially of Plato and Aristotle. These sources, while differing in many ways, agreed in one fundamental aspect which differs radically from both indigenous and some eastern attitudes: rather than seeing humans and nature as existing in some kind of kinship or partnership, the west has tended to believe that nature exists primarily to serve human needs. Aristotle (1967, Book 1, Chapter 5) said:

Plants exist for the sake of animals, and animals exist for the sake of man.
This notion was still alive twenty-one centuries later, when Kant (1963) said:
Animals are...here merely as a means to an end. That end is man.

The ancient notion that humans are distinct from and superior to nature was, then, deeply entrenched in western thinking and fed into the concepts of nature employed by science and philosophy in the early modern period (Thomas, 1983; Regan and Singer, 1989, pp. 60–130) and retain a strong influence on modern thinking (Passmore, 1974, Chapter 1).

Western culture is unique among the world's cultures in that all of its sources share fundamental assumptions about the value of nature (Thomas, 1983 pp. 17–25; White, 1967). That conception has come to be characterised as *anthropocentrism* or human-centredness. This traditional basis of western environmental ethics has come under attack in recent years but remains the most common and deeply entrenched set of attitudes regarding the environment in the western world. Despite its limitations it is a perfectly adequate framework for many common cases of environmental decision making. For those reasons it is important to understand its general outlines.

Its major characteristic is that it sees the value of nature in human terms. For the extreme anthropocentrist, the value of nature resides wholly in its *usefulness* in furthering human well-being. To have an *anthropocentric* attitude to nature does not mean simply to see things from a human point of view but rather to believe that humans and their interests are the most important things in existence. Thus, an anthropocentrist believes that humans or some characteristic of humans, such as reason or happiness, are entities of higher value than others and that nature has value only in so far as it furthers these things, or at least that this is the only rational basis for environmental decision-making (Passmore, 1974, Chapter 7). This has been the traditional view of the western world and it has held sway into the 21st century.

Figure 4.1 summarises its main assumptions that have traditionally underpinned the western world's view of the human relation, and source of obligation to, the natural world (Passmore, 1974).

Figure 4.2 presents an example of an anthropocentric ethical argument. As far as it goes, it is a compelling (if somewhat simplistic and unrealistic) argument. Unless you were confronted by an immoral corporation, it would be effective in changing their behaviour. Even if the corporation was immoral, this style of argument is likely to convince enough people to make laws against such behaviour effective. There are, then, many cases where the traditional human-centred view supplies

---

• Humans are the planet's most important and most valuable species;
• Humans are distinct from and superior to the rest of nature (in some respect);
• Humans have sufficient skill and knowledge to manage nature ('Humans know best'); and
• The value of nature is its instrumental value (nature is 'good-for-us' not 'good-in-itself').

Figure 4.1 Anthropocentric assumptions

Imagine a corporation decides, for the sake of profitability, to dump some toxic waste into a nearby creek. An anthropocentric argument against this practice might be: it is generally regarded as immoral to poison or otherwise harm one's neighbours. This practise will almost certainly harm your corporation's neighbours. If you dumped this toxin directly into your neighbour's water supply, you would rightly be held to be an environmental vandal (or worse, you could be prosecuted). Therefore you ought to desist.

Figure 4.2 Anthropocentric ethical argument

adequate reasons to protect the environment. But this approach has its limitations. There may be cases where the waste substance is not directly harmful to humans but damages the creek ecosystem in some way. In that case, this framework loses much of its force. More importantly, this approach invites trade-offs. If the gain in terms of wealth creation is great and the damage to the ecosystem not too serious, the community might decide (indeed has decided in many such cases) that the benefit outweighs the cost. It is this sort of concern – that in the anthropocentric framework the environment is not directly valued for its own sake, and is therefore vulnerable – that has encouraged the emergence of an alternative ethical framework.

4.7 Ecocentrism

In recent times many people have found the anthropocentric framework inadequate as an explanation of their reasons for wanting the environment protected, and this has prompted ethical theorists to construct non-anthropocentric theories (Callicott, 1994). It has commonly been argued that as a minimum the anthropocentric approach is too narrow (Norton, 1995) and, as a maximum, it is one of the major reasons underlying environmental degradation (Callicott, 1999).

Therefore a broad strand of thought has emerged as a challenge to the anthropocentric view. It has become variously known as biocentrism (Taylor, 1981), more commonly as *ecocentrism* (Callicott, 1984) and more popularly as Deep Ecology (Fox, 1995). Ecocentrism derives mainly from the thoughts of Leopold (1987) and Naess (1973). Leopold argued that human beings should no longer see themselves as separate from the land but rather as 'members and plain citizens' of it, with no privileged place. In a famous statement of an ethical principle, he argued that the rightness of an action should no longer be judged by such considerations as the amount of human well-being it brings about or considerations of the common good or individual rights, but an action is right:

...when it tends to preserve the integrity, stability and beauty of the biotic community. It is wrong if it tends otherwise.

While this principle and allied ideas have been criticised by ethical theorists (Regan, 1984; Rollin, 1993; Paden, 1990), the general attitude and orientation to nature that they attempt to express has found a deep resonance in many sections of the community. Underlying the anthropocentric attitude is the idea that nature's value lies ultimately in its usefulness to humans. Underlying ideas like Leopold's and Naess' is the notion that as well as this instrumental value, nature has intrinsic value i.e. a value in its own right independent of its relation to human beings. While ecocentrists differ widely in their interpretations, the central assumptions are summed up in Figure 4.3.

These assumptions constitute a major challenge to the anthropocentric view and they are gaining widespread influence in western society. While there is no clear consensus among environmental ethicists about the meaning or the implications of these ideas (Paden, 1990; Rollin, 1993), intrinsic value often appears in the list of values to be protected in environmental legislation. Public discussions of environmental problems suggests that the community is willing to overlook logical or conceptual problems raised by academic critics and embrace at least the general non-anthropocentric tendency of this style of thought. It is therefore be prudent for any environmentally concerned corporation to be familiar with ecocentric ideas. Figure 4.4 presents an example of an ecocentric ethical argument.

It will be noticed that the conclusions of the two over-simplified hypothetical examples of Figures 4.2 and 4.4 are the same. But it is important to notice that this will not always be the case. Sometimes, the anthropocentric argument will give a different outcome to the ecocentric one. It then becomes a matter of judgment to decide when to apply the appropriate framework (though, ecocentric critics of anthropocentrism will deny this and claim that ecocentrism in some form is always more appropriate, and vice versa). The argument above depends on the acceptance of the notion of nature's *intrinsic value,* which is by no means universal. Future generations will probably judge how well corporate managers have resolved this argument.

- Humans are just one species among many with no special privilege;
- Humans are an integral part of the land (Leopold) or the 'web of life' (Naess);
- Humans do not have sufficient skill or knowledge of ecosystem function to manage the environment ('Nature knows best'); and
- Nature has intrinsic value (is not only 'good-for-us' but is also 'good-in-itself').

Figure 4.3   Ecocentrism assumptions

Imagine a corporation decides, for the sake of profitability, to dump some toxic waste into a nearby creek. An ecocentric argument against this practice might be: it is generally agreed that a creek ecosystem has intrinsic value. We ought never to harm anything that is intrinsically valuable without strong justification. Increased profit is not sufficient justification. Therefore you ought to desist.

Figure 4.4   Ecocentric ethical argument

4.8 Sustainability

Because of these kinds of conflicts, alternative frameworks have arisen, most notably, the notion of *sustainability*.

The meaning of the term *sustainability* has undergone a fundamental shift in recent years. From its original meaning of 'exploiting the maximum yield consistent with maintaining the stock of renewable resources' such as for a fishery, it has taken on a broader meaning and has been transformed from 'sustainable yield' to 'sustainable development'. This shift is from a fairly clear and pragmatic concept to a vaguer one, possibly so vague that it is in danger of becoming a meaningless motherhood concept: 'Nobody opposes it because nobody knows what it is' (Norton, 1992). While there is some force in this view, the general intent is clear enough, and if taken seriously, places further constraints on corporate behaviour.

Corporations are obliged to comply with laws and regulations, including complex environmentally-oriented ones that have grown exponentially in recent decades, in response to community demands and expectations. In this new climate, Board members and CEOs can find themselves liable for large fines or imprisonment for actions (that breach environmental legislation) which a generation ago would have been considered normal business practice.

It is little wonder, that many corporate executives feel uncomfortable when confronted with ideas that together suggest they have responsibilities beyond mere compliance with laws and which take them into the uncharted territory suggested by the vague term *sustainability*.

Values underlying sustainability

Sustainability and *sustainable development* have become umbrella terms for environmentally and socially responsible behaviour. But what is sustainability? Brundtland (1987) defines sustainable development as:

> ...development that meets the needs of the present without compromising the ability of future generations to meet their own needs.

This definition, with its apparent focus on human well-being, seems squarely human-centred. The ability to meet the needs of future generations implies preserving nature in a sufficiently healthy state to allow such needs to be met. It can therefore be seen as a compromise between the anthropocentric and ecocentric frameworks. Notwithstanding, it also embodies certain values including the concept of *needs* both of present and future people. In other words, the principle of sustainable development seeks to extend the anthropocentric idea of meeting *human needs* for the current generation (intra-generational equity) to meeting human needs of future generations (inter-generational equity).

Intra-generational equity embodies the idea that all people currently alive have an equal right to access to the means of securing a decent life (however that is defined) and that no group has a right to deprive any other group of such access.

Inter-generational equity embodies the idea that the human community is not merely a temporal succession of distinct groups, but rather is a partnership between generations. Therefore inter-generational equity is a belief that extends the notion of social and environmental justice into the future.

Now we cannot know with any certainty what the *specific needs* of future people will be, but we do know that they will need at least the same basics of life as ourselves and sufficient natural resources to have a range development options. (For sustainability these may be less than currently enjoyed by western countries.) So sustainable development (Brundtland, 1987) suggests as a minimum we have an obligation to use resources to fulfill our own needs in such a way that natural processes on which these things depend are not diminished.

To support these societal values, a corporation would need to acknowledge that its operations should be seen in the light of their effects on nature and current and future societies. But why should a corporation be involved with sustainable development at all? One important reason is that the idea of sustainable development has become the rallying point not merely for governments and environmental activist groups but also for large sections of the community.

4.9 Summary

This chapter has discussed some of the ideas and values that underpin two different belief systems (anthropocentrism and ecocentrism) by drawing on the environmental perspectives of two cultural extremes – indigenous and western. An argument has been developed to link sustainability to these belief systems and to explain other values that underpin the concept of sustainability. Before business organisations (corporations) embrace sustainability they should at least understand its ethical basis.

5
Politics and the Environment

Ros Taplin

5.1 Introduction

This chapter discusses the different political influences and constraints that can affect management of the environment. These include the: timescales of politics, electoral cycle and environmental change; role of scientific knowledge; organisation of the functions of government; government and business relationships; dichotomies between the economic paradigm and the environmental paradigm; and environmental governance.

Protecting the environment is one of the central functions of the contemporary nation state. Environmental issues now assume an important role in the political decision making sphere at local, national and international levels. However, this responsibility has only evolved as a significant political concern of governments since the late 1960s. The traditional roles of the nation state include law and order, security, defence, health, economy, taxation, welfare, education and at the international level, foreign relations including strategic and trade relationships between sovereign states. Since representatives of governments met in Stockholm in 1972 at the inaugural United Nations (UN) Conference on the Human Environment, the environment has become a further responsibility of national governments and also the international community as a whole to oversee and manage. As a corollary to this, the performance of governments on environmental issues has become important to the voting public in democratic nations.

5.2 Timescale of politics and environmental change

The timescale of politics is very different from that of environmental change e.g. the changes projected for the next 50 to 100 years due to global warming are insignificant in comparison to the history of climate change on Earth but in comparison to the timescales of politics, elections and campaign promises, political parties, strategic leadership and planning, greenhouse impacts are a very long term issue. John Gordon (1996, p. 142) who spent many years in government in the United Kingdom (UK) has reflected that:

> Politics in Britain and most other democratic countries is overwhelmingly centred around short-term issues and two- to five-year electoral cycles. Politicians need to be re-elected to survive and by the very nature of the traditional political process everything is subordinated to this interest...short-term issues of recession, unemployment and crime at home and regional conflict abroad inevitably take priority over addressing long-term issues such as global pollution, resource depletion, and population increase.

In a democracy, once a government has been elected, inevitably, there are long-term planning issues that have to be addressed. Major land-use, industrial, commercial and infrastructure developments such as manufacturing plants, power stations, dams, roads, bridges and urban residential developments, require decisions and continue in use for several decades if not longer. Since the 1970s, political decision-makers on these issues as well as shorter-term imperatives have had to grapple with environmental as well as economic considerations. Environmental impact assessment (EIA) (Chapter 25) was required in the United States of America (US) from the inception of the National Environmental Protection Agency in 1970 and other nations followed in introducing mandatory EIAs.

The Brundtland Report addressed the political economic-environmental dichotomies and timescale considerations with recommendations that environmental protection and sustainable development concerns (development that meets the need of the present without compromising the ability of future generations to meet their needs) were important and were the responsibility of all national governments (WCED, 1987).

In the early 1990s opinion polls showed that an increasing fraction of the public was becoming concerned about long-term issues (Lowe, 1994, p. 318). Gordon (1996) however was pessimistic about governments' ability to address environmental concerns in a responsible manner while balancing electoral pressures and timescales with environmental needs. He talked about 'failure of timescales' both at national and international levels. In particular, a failure of timescale has become apparent with regard to the issue of climate change (Taplin, 1996). Even though it is over a decade after the Framework Convention on Climate Change was opened for signing and almost two decades since scientists first expressed apprehension about the enhanced greenhouse effect, the issue has not yet been seriously addressed by nations such as the United States, Australia and the Russian Federation either via domestic policy or via ratification of the Kyoto Protocol. This has been due to short-term economic interests, notwithstanding the long-term impacts predicted.

5.3 Institutions involved in environmental protection

The institutions of the nation state that are involved in management of the environment, or in critiquing the management process, virtually cover the entire organisational scope of a nation (Davis et al, 1993a, p. 17) and include:

- the constitution, elected representatives or politicians (including ministers, cabinets) parliament or congress;
- non-elected administrative departments at the national level and their policies;
- legislation, the judiciary and the courts;
- enforcement of public order and defence of territory;
- sub-national apparatuses such as provincial/state government and local government/municipal authorities;
- regulatory agencies such as those directed towards environmental protection and development or planning control;
- state trading authorities such as water utilities and electricity authorities (though privatisation has moved many of these bodies into the commercial arena in the last decade in the UK and Australia);
- semi-state institutions such as political parties and trade unions; and
- in a broader sense, non-government organisations (NGO), the media, the establishment churches and even the creative arts.

In the past these institutions were primarily focused on economic prosperity and social justice. However, in addressing environmental concerns in the past three decades, these institutions have undergone transformation 'to achieve a new synthesis between apparently conflicting (economic and environmental) approaches' (Papadakis 1993, p. 104).

5.4 Political influences on environmental decision making

There are a variety of political interests or players who influence environmental outcomes in democratic societies including voter influence, party influence, media influence, lobbying of special interest groups, and the contribution of researchers and knowledge.

Political parties

Traditionally there have been socio-economic bases of support for political parties. Dalton and Kuechler (1990) and Inglehart (1990) have claimed that the 'new' politics that has evolved due to environmental concerns has undermined the traditional class based party divisions. Papadakis (1993) however contends that: 'It is uncertain whether groups concerned with issues like the environment can bring about a long-term shift in loyalties from established political parties...the influence of environmentalism implies not so much a realignment as a *de*alignment of voters...it has not led to new clearly identifiable divisions between major parties but to the weakening of traditional political alignments.'

Green parties have arisen in many countries (e.g. Germany, Australia, and UK) as a reaction to the traditional right-left concerns of political parties and their lack of accommodation of the environment as a central concern. The degree of success of these parties waxes and wanes and has been related to the structural nature of national electoral systems as well as to voter allegiances.

Media

The media have a very significant role as gatekeepers in bringing environmental or sustainability issues onto the public agenda or conversely suppressing them. Papadakis (1993) reflects that 'Environmentalists have become highly skilled in gaining publicity. The media in turn has profited from public interest in the environment.' Conversely, Lowe (1994, p. 321) criticises the media as being supportive of economic rationalism rather than environmental concerns.

Interest groups

A further significant influence is that of interest groups. Environment and business NGOs play an important role in shaping environmental policy outcomes and have become increasingly sophisticated in their acquisition and use of scientific and economic information in lobbying government (Weale, 1992). Davis et al (1993a) say in relation to Australia that '...groups do not speak with equal voices. Their uneven influence is a cause, and a reflection, of the distribution of power within Australian society. Some parts of society cannot organise at all.' This is equally evidenced in other democracies. Doyle and Kellow (1995, p. 121), however, points out that it is not always major interest groups that have the most influence over environmental outcomes saying 'The relative power of interest groups is not a fixed equation...and relatively powerless groups are capable of transforming the balance.'

Epistemic communities

Another important influence from the 1990s onwards on environmental politics and policy making has been the contribution of environmental scientists, economists and bureaucrats working together as 'epistemic communities' or '...transnational networks of knowledge-based communities that are both politically empowered through their claims to exercise authoritative knowledge and motivated by shared causal and principled beliefs...' (Hass, 1992). They play an important role in agenda setting and in particular have had significant political impact e.g. on agreements for the Mediterranean, ozone, and climate change. They influence problem definition which may involve more than one scientific discipline, but if more than one discipline is involved, then they work together sharing a common body of facts and also interpreting information in a similar way (Hass, 1992).

5.5 Political discourse and sustainable development

Political rhetoric in relation to the environment in the 1990s and 2000s has centred on dialogue on 'sustainability' and 'sustainable development.' These terms have been highly disputed in terms in their definition arguably due to the 'political visions' they convey. Varying definitions have been given by different writers. Meadows et al (1992) state that 'A sustainable society is one that can persist over generations, one that is far-seeing enough, flexible enough, and wise enough not to undermine either its physical or social systems of support.' In contrast, O'Riordan and Voisey (1998) say that 'The sustainability transition is the process of coming to terms with sustainability in all its deeply rich ecological, social, ethical and economic dimensions...It is about new ways of knowing, of being differently human in a threatened but cooperative world...'. Birkeland (2002) gives a further definition of sustainability as '...the ability of future generations to achieve the same level of natural, social and cultural resources enjoyed by the current generation.'

Going further than definitions, the terms that are regularly used in discussion of sustainability are: sustainability indicators; the precautionary principle; the polluter pays principle; intergenerational and intragenerational equity; staying within source and sink constraints; and maintenance of natural capital at or near current levels.

Because of the variations in perspective, many challenges have been made that sustainability is too vague (Young, 1992). Hajer (1995), however, sees sustainable development as a political discourse and refers to 'discourse coalitions' as groups who view environmental politics in a relatively uniform manner. He says: '...sustainable development...is much more a struggle between various unconventional political coalitions, each made up of such actors as scientists, politicians, activists, or organizations representing such actors, but also having links with specific television channels, journal and newspapers or even celebrities.'

Discourses on sustainable development have achieved public and political prominence, and have had an impact on political decision making. Dryzek (1997, p. 199) refers to the multiplicity of actors involved in the sustainability dialogue: 'While at first glance the sheer variety of available definitions of sustainable development seems like a defect of this discourse, from the perspective of social learning it is a distinct advantage, for it does not rule out a variety of experiments in what sustainability can mean in different contexts, including the global context.' He also reflects that: 'There is still no consensus on the exact meaning of sustainability; but sustainability is the axis around which discussion occurs....'

In contrast, Fien and Tilbury (2002) holds the perspective that '...those that argue that sustainable development is too ambiguous a concept... have missed the importance of sustainable development as a way of transcending conflicting worldviews'; and that it facilitates different stakeholders with different values and beliefs coming together. Fein points out that the definition of the term has been negotiated at the UN level since Rio and that education has become to be seen as an essential component the process moving towards sustainability. *From Rio to Johannesburg* (UNESCO, 2002, p. 7) says: 'Since 1992, an international consensus has emerged that achieving sustainable development is essentially a process of learning.'

5.6 Paradigm shift

Birkeland (2002) argues that reconciliation of growth and the environment '...requires paradigm shift in...the organizational culture of business and industry.' This comment relates to an ongoing academic exchange of ideas on political and social paradigms and their impact on the environment. A paradigm is defined by Milbrath (1989, p. 116) as '...a society's dominant belief structure that organizes the way people perceive and interpret the functioning of the world around them.' Table 5.1 shows Milbrath's delineation of paradigms.

Table 5.1   Contrasts between economic and environmental paradigms

Economic Paradigm	Environmental Paradigm
Risk acceptable to maximise wealth	Careful plans and actions to avoid risk
Economic growth unlimited No resource shortages	Limits to growth Resource shortages
Large projects	Quality of life and simpler lifestyles
Domination of nature Avid consumption	High valuation on nature
Competition Hierarchy	Cooperation Consultation and participation
Compassion only for close friends and family	Compassion for other people and generations and compassion towards other species

Source: Modified from Milbrath (1989, p. 119).

According to Papadakis (1993) most environmental actors hold halfway positions between these two paradigms. He says a strong reform element is evident in environmental movements and intermediate positions are held by most participants. He says: 'Only a small core of participants in social movements make sharp distinctions between green and other values. The majority tend to borrow from both paradigms and from different ethical positions. Even if they do not make sharp distinctions between their own and established values, they are often prepared to negotiate and strike bargains.' He also reflects 'The emergence of intermediate positions (in other words positions that have drawn on both a dominant economic and an alternative environmental paradigm) undermines the claims of many commentators...that environmentalism has replaced socialism as the enemy of capitalist development.'

5.7 Government-business relationships and ecological modernisation

Business organisations have become increasingly conscious of the environmental consequences of their operations and the implications for their strategic direction. Since the early 1990s much discussion of the relationship between business and government has focussed on *ecological modernisation* or industrial transformation (Anderson and Massa, 2000; Christoff, 1996; Gibbs, 2000; Hajer, 1995; Jordan et al, 2002, 2003a, 2003b; Knill and Lenschow, 2000). This perspective concentrates on the contemporary evolution of interactions between environment, economy, society and the political arena. Government imposed 'command and control' measures to manage polluters are seen to be 'old' instruments that will be phased out with a shift to self-regulation and incentives for industry. An intermediary phase of co-existence of 'new' and 'old' policy instruments together with 'new' and 'old' institutional structures to facilitate the implementation of agreements is expected. The theoretical assumption regarding so-called new environmental policy instruments (NEPIs) is that they have greater effectiveness and efficiency.

The concept that such a change in the role of government would have positive environmental outcomes was initially academic but has been embraced by politicians and, in particular, in European nations where *ecological modernisation* has been underway for nearly two decades (Weale, 1992). UK Prime Minister, Tony Blair (2003) said: 'The challenge for this Government today is in continuing to integrate the goal of environmental modernisation into our vision of Britain. To combine greater economic development with better environmental impact – bringing the environment, economic development and social justice together.'

Driving forces

Jordan et al (2003a) identify several *driving forces* for the implementation of ecological modernisation: dissatisfaction with environmental regulation;

perceived superiority of NEPIs; neo-liberal ideas about deregulation; influence of the European Union (EU) in forcing adoption of EPIs by some member states; growing international competition and economic recession driving a search for more cost-effective policy instruments; growing domestic support from elected politicians; interaction between different types of NEPI (e.g. the threat of environmental taxes has popularised voluntary agreements and emissions trading); and climate change has pushed governments to actively explore emissions trading and voluntary agreements.

From the early 1990s onwards, *ecological modernisation* perspectives spread beyond Europe via environmental regimes. The WCED (1987) report, the UN conferences in Rio in 1992 and Johannesburg in 2002, and the influence of international organisations (UNEP, OECD and the Global Ecolabelling Network) have strongly shaped and influenced the direction of environmental politics. As a result of this, in many nations micro-scale (voluntary agreements, eco-labelling, environmental and carbon taxes, and emissions trading) and meso-scale initiatives (national environmental plans and strategies for sustainable development) have been instituted. In particular, implementation of the 1992 Framework Convention on Climate Change has been characterised by adoption of voluntary agreements.

Tews et al (2001) observed: 'The recent shift in the prevailing policy pattern from a sectorally fragmented and largely legal regulatory mode to an integrated environmental policy approach relying increasingly on "softer" and/or more flexible instruments such as voluntary instruments, ecolabels, or ecological tax reforms is equally proceeding on a global scale.'

Barriers

Jordan et al (2003a) have reported that *barriers* to introduction of ecological modernisation in Europe have included: opposition from environmental pressure groups; opposition from vested interests (e.g. energy intensive industry opposition to carbon taxes and emissions trading); fears about competitiveness; public protests related to distributional impacts of environmental taxes (i.e. related to rising fuel prices); and EU requirements (e.g. voluntary agreements taking the form of legal contracts). Similar barriers have occurred in other nations (Taplin, 2003).

5.8   Towards environmental governance

Arguably, many nations have moved some way since the 1960s towards *environmental governance* systems that place an important focus on environmental concerns and are not predominantly centred in the economic paradigm. Environmental governance refers to the written and unwritten policies, decision making, and behaviour that affect the way the environment is managed – particularly with regard to openness, participation, accountability and effectiveness.

Milbrath (1989, p. 282, 301–02) has proposed overlaying a *learning governance structure* on a traditional nation state's structure with the potential of strengthening government in the move towards environmental governance and a sustainable society. He recommends that a governance structure should have four basic components:

- an education and information system;
- a systemic and futures thinking capacity;
- an intervention capability for stakeholders; and
- a long range sustainability impact assessment capacity for every policy initiative.

The advantages of this structure are that governments that can better anticipate the future can have a better chance of: *dealing* with such problems as climate change, overpopulation, resource shortages, species extinction and ecosystem damage; and *undertaking* the new functions that are being assigned to it e.g. encouraging social learning, facilitating quality of life, and helping society to become sustainable.

Many government functions and undertakings that have emerged in the last decade do have parallels with the components of Milbrath's governance structure. Nevertheless, as Dryzek (1997) comments: 'Environmentalism already flourishes in opposition to industrialism; but much remains to be done if industrial society is ever to give way to ecological society.'

5.9 Summary

This chapter has presented some of the political issues that can affect the management of the environment by a nation state. It has discussed the disparity in time scales between politics and the environment, the institutions involved in the political process and how they can affect government decision making. Political and institutional changes are necessary to make the paradigm shift to a sustainable society. These changes are occurring now and business organisations need to respond proactively to them.

5.10 Further reading

Elliott (1998); Kanie and Hass (2004); Miller (1995); Sterling (2001); Walker (1994); Young (1997).

5.11 Questions

- Select an environmental case (e.g. a dispute over forestry, genetically modified crops, climate change). Describe its nature and analyse the political processes involved and the role of various stakeholders and political influences e.g. NGOs, general community involvement,

industry lobbying, scientists or researchers and the media. How did politics influence the environmental outcomes?
- Consider Milbrath's governance framework (Section 5.8) and evaluate current government performance on the environment in your nation i.e. environmental education and environmental information resources; futures thinking capability and government agency involvement; facilitation of stakeholder's intervention capacity; and strategic environmental assessment of all policy initiatives.

6
Government Environmental Decisions

Robert Staib

6.1 Introduction

This chapter discusses some of the issues associated with how governments manage the environment. Government includes government ministers, their departments, agencies and independent authorities i.e. the public service or the bureaucracy. The objective is not to provide a detailed description of government environmental management but to introduce the reader to some ways governments organise and manage environmental aspects and some of the approaches they use in making environmental decisions.

Business organisations need to communicate and liaise with governments in a variety of ways and at different levels. They need to know how the bureaucracy functions, who administers which portfolio and department, how legislation and policy are applied, how environmental decisions are made and what changes to these aspects are likely in the future. Table 6.1 lists some of these aspects and indicates the aspects discussed in this chapter and others.

Table 6.1 Environmental aspects for government environmental management

Category	Aspects	Examples	*
Bureaucracy	Ministers	Minister for Environment	6
	Departments/agencies	Environment Protection Agency	6
	Partnerships	Intergovernmental	6
Drivers	Government policies	On uranium mining	5
	Law	Clean air legislation	7
	Public opinion and voters	Protests, elections,	3,5
	Interest groups	Green groups, consumer groups	9,19
	Agreements	Intergovernmental and international	6
	Agencies	Intergovernmental and international	6

Table 6.1 Environmental aspects for government environmental management
– *continued*

Category	Aspects	Examples	*
Structure	Governments.	National, regional, local	6
	Government agencies:		6
	With environ. portfolios	Environmental protection, national parks, land & water conservation,	
	With environ. branches	sustainable energy, flora/ fauna.	
		Urban planning, fisheries, waterways,	
	Which can influence	forestry, water & sewage, health.	
		Transport, industry, treasury, trade	
Decisions	Policy, program, project	Cost benefit analysis	6
	Urban plans, projects	Strategic environmental assessment	6,25
Planning	Urban planning	Policies, regional plans, local plans,	–
	Building developments	Development approvals	25
Implementation	Legislation	Licences, penalties, monitoring	7
	Economic instruments	Tradeable permits	8
	Voluntary agreements	Incentives, offsets	8
	Govt. supply contracts	Environmental conditions	–
Standards	Environ. mgt. systems	For government supply contracts	12
	Building standards	For energy, materials	24
	Emission standards	For water, air, land, waste	–
Education	Training	Internal and for industry	–
	Measurement/reporting,	State of the environment reports	–
	Information provision	Endangered species	–
	Industry development	Sustainable energy grants	
	Universities	Environmental management courses	–

*Refer to chapter listed for further discussion.

6.2 The government environmental bureaucracy

A simple view is that the elected Government makes decisions and defines policy initiatives in a form in which they will be achieved, then raises revenue and distributes resources through programs and agencies i.e. the bureaucracy administers decisions made by the elected Government (Doyle and McEachern, 1998, p. 149).

It is obviously more complex than this with the bureaucracy providing advice, information and assessments back to Government to help it establish policies and to make decisions. The Government is also affected by perceived and measured public opinion (Chapter 3) and through the political processes (Chapter 5). Environmental issues and decisions can also take place in atmosphere of conflict between stakeholders e.g. uranium mining, forest clearing and dam building (Doyle and McEachern, 1998, p. 159).

Prior to the establishing of environmental protection agencies in the late 1960s and early 1970s, governmental environmental responsibilities for pollution in many countries lay with other departments e.g. soil conservation (Eccleston, 1999, p. 14), health, maritime services, sewage treatment (Staib, 1997, p. 8). With the evidence of increasing pollution, new legislation was passed and separate environmental protection agencies were established to administer it. This provided a better focus on environmental issues e.g. National Environmental Policy Act (NEPA) in the USA. Many countries have established an environmental protection agency/authority (EPA), though in variously different forms (Eccleston, 1999; Section 6.6).

Government structure

Over the years other specialist Government environmental agencies and departments have been established as well as environmental branches within other government administrative areas e.g. fisheries, waterways, national parks, urban planning, water conservation, forestry, water treatment and supply. For some departments environment is a major focus while with others it is a small part of their main responsibility to provide a service or manage state owned assets. It is often difficult to fit some of the environmental issues discretely into individual departments e.g. air pollution can extend over departments like transport, industry, environmental protection (Section 6.6).

The fragmentation of departments and environmental responsibilities can lead to duplication and lack of coordination. One response is to amalgamate departments and portfolios to form large units e.g. in NSW Australia, a response has been to try and improve coordination of environmental issues by creating mega environmental departments by drawing previously separate portfolios into the one department under a single government Minister e.g. planning, land and water conservation into the one department and environmental protection (ex. EPA) and national parks into another department and Minister.

On the other hand specialist administrative functions and departments are often necessary to deliver these programs. It is therefore necessary to divide the programs/areas of government into bite-sized chunks and establish a hierarchy. There may always be conflict between departments but it is important to try to minimise overlapping of responsibilities and achieve coordination (Beale, 1995).

Whole of government approach

Yencken (2002) believes that single portfolio (or specialist environmental portfolios) response to environmental issues by government is no longer enough to respond to current environmental issues. Environmental issues are linked to both social and economic issues and can extend across many non-environmental portfolios e.g. health, social equity, human settlement, transport and consumer issues. Opportunities for economic transformation of, and approaches to environmental issues belong to economic portfolios of government e.g. treasury, industry, trade etc. Environment can no longer be sectoralised. It is not the province of one minister or department or even one country and there needs to be an integrated approach: a whole-of-government approach. This whole-of-government requires legislative change, government wide monitoring, an extensive information base and coordinated environmental actions of all government departments (Yencken, 2002).

There is a need to balance environmental issues within a social and economic framework. Many countries have a hierarchical governmental system with federal/national, state/county/regional and local government/council level with varying environmental powers. The national level is important for the implementation of international agreements and commitments at the national level while the local government is representative of the daily lives of the community (Section 6.6). There is increasingly a need for strong intergovernmental arrangements both within countries and internationally. Some examples are discussed in Section 6.4.

6.3 Government environmental decisions

Government departments and agencies become involved in assessing the value of proceeding with or continuing Government actions e.g. policies, programs, provision of services or products. Economic assessment comprises a central part of any assessment but many agencies are required to routinely include both environmental and social issues alongside the economic assessment. The requirement and the extent of these assessments may be required by legislation or could be defined in an agency's management process. There is a hierarchy down which these could be required starting with legislation and proceeding through regulations, policies, programs and ending with specific projects or ongoing operations. E.g. a regulatory impact statement (supported by a cost-benefit analysis which included environmental issues) was prepared in 1995 prior to Sydney Australia introducing a Regulation to control burning of refuse by households in an attempt to reduce air pollution (Aquatec Environmental Consultants, 1996).

Cost benefit analysis

Commonly used assessment techniques include cost benefit analysis (CBA) and cost effectiveness analysis (CEA). Applying CBA requires quantification of all environmental and social benefits and costs by using various techniques (Chapter 8, 14) and balancing these against economic values. It can be used as an absolute measurement or to compare alternatives. CEA does not generally require calculation of benefits but calculates the least costly way of achieving of a previous agreed objective e.g. Sydney's water supply authority (a Government owned organisation) used CEA to determine the most economical way to meet its agreed objective of improving the quality of the city's drinking water (Aquatec Environmental Consultants, 1996).

While CBA can be used at any stage of the above hierarchy, more information is generally available at the project stage where much environmental information is gathered to support environmental impact assessment processes (Chapter 25). CBA is not an absolute decision making tool but is used by government agencies to support their decisions. It should always be subject to a sensitivity analysis by varying the values (or assumptions) attributed to environmental data that is either critical to the decision or that has a low level of accuracy (Harding, 1998, p. 145).

Strategic environmental assessment

Strategic environmental assessment (SEA) is environmental assessment applied at the level of government policies, plans and programs. It can be applied at different levels e.g. sectorial (waste disposal, energy, forestry, transportation); regional (region, city or local area); and indirect (science and technology, financial, justice, immigration) (Court et al, 1996). Applied at the national government level and spreading down through the state/ regional government to local government it can set the framework for better environmental performance of a nation state. There are many examples of how governments have attempted to integrate environmental assessments into their governance e.g. the United Kingdom Cabinet requires a Regulatory Impact Statement with the presentation of every new regulation and this includes a requirement to consider environmental issues (Regulatory Impact Unit, 2003).

Environmental planning undertaken by public urban planning authorities is a form of SEA though it is probably not sufficient in itself without more detailed environmental impact assessment (Chapter 25) to enable the environment to be fully protected (Thomas, 2001, p. 212; Staib, 2003a).

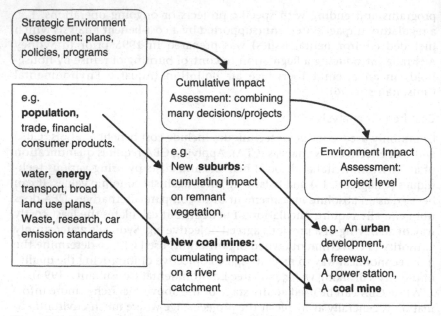

Figure 6.1   Environmental assessment framework
Source: Based on Thomas (1998, p. 47, p. 212); Court et al (1996).

USA's National Environmental Policy Act 1969 was one of the first pieces of legislation requiring Government to assess the impact of its actions. It requires 'federal agencies to consider environmental issues in reaching decisions ....' (Eccleston, 1999, p. 11), which depending on the size of the project may require the production of an Environmental Impact Statement or an Environmental Review. Many USA agencies have used the process to facilitate planning e.g. production of Resource Management Plans by the Bureau of Land Management (Eccleston, 1999, p. 59).

Figure 6.1 illustrates the concept of a hierarchical approach from SEA to cumulative impact assessment (CIA) to project level environmental impact assessment (EIA). For complete adoption it requires a considerable governmental organisation and the collection and maintaining of much environmental data. A lot of this data is now being developed in many countries with State of the Environment Reporting.

Cumulative impact assessment

Although in larger EIAs private organisations are often required to undertake a limited amount of CIA, it is mainly the province of governments. E.g. CIA is becoming a larger part of urban planning in Sydney Australia where a hierarchy of planning instruments are in place, starting with state and regional planning policies and plans moving through local environmental plans and then to development control plans for individual precincts. Recently the State national parks and wildlife department undertook a Sydney region wide remnant vegetation study and assessment to make it easier to assess the significance of smaller though cumulating impacts as new suburbs are approved and individual precincts are built. EIA is discussed in Chapter 25.

6.4 Intergovernmental partnerships

Many partnerships are being developed within countries and between countries to help address major environmental issues especially those issues that extend beyond local boundaries and beyond individual countries. A sample of these partnerships is summarised below.

A major international partnership is the Montreal Protocol on Substances that Deplete the Ozone Layer. It was adopted in September 1987 and is an international agreement to phase out ozone depleting substances. It is supported by a fund that finances incremental costs incurred by developing countries in phasing out their consumption and production of ozone-depleting substances (United Nations, 2004a). Another well-known international agreement is the United Nations framework convention on climate change called the Kyoto Protocol. 'The convention sets an overall framework for intergovernmental efforts to tackle climate change. It establishes an objective and principles and spells out commitments for different groups of countries according to their circumstances and needs.' As of May 2004 it has been ratified by 189 countries (UNFCCC, 2003).

The European Union is a partnership of 25 separate countries supported by a European Parliament and several other institutions. All European Union (EU) decisions and procedures are based on Treaties, which are agreed by all the EU countries. These include environmental matters. The range of environmental instruments used has expanded as environmental policy has developed and include: framework legislation providing for a high level of environmental protection while guaranteeing the operation of the internal market; a financial instrument to co-finance environmental projects; and technical instruments e.g. eco-labelling, the Community system of environmental management and auditing, a system for assessment of the effects of public and private projects on the environment, and the criteria applicable to environmental inspections in the Member States (European Union, 2004).

At a national level, the Australian Commonwealth government has environmental partnerships with the Australian States on environmental matters. The National Environmental Protection Council brings together the Environmental Protection Authorities of each Australian state on matters of emissions and waste to air, water and land. The Australian and

New Zealand Environmental Conservation Council consist of federal and state ministers from both countries supported by a standing committee of officials and is supported by an Inter-Governmental Agreement on the Environment (Yencken, 2002).

6.5   Summary: what business needs to do

'Whilst environmental decisions are informed by scientific and technical inputs, the institutional and political framework in which they occur is the primary determinant of outcomes. Structures of government, its agencies, advisory bodies, legislation and processes for decision-making determine the relative rights and potential influence of various stakeholders in the decision-making process' (Harding, 1998, p. 255).

Business organisations need to: communicate and liaise with governments in a variety of ways and at different levels; know how the bureaucracy functions, who administers which portfolio and department, how legislation and policy are applied and what changes in all these aspects are likely in the future. The framework in which they operate their business needs to take into account not only government legislation but also environmental regulation and policies some of which have and are coming from intergovernmental agreements including an increasing number from international agreements. From a strategic planning point of view organisations need to anticipate and be in a position to respond future developments in this area.

In this chapter we have briefly discussed some of the environmental issues associated with governments and referred to other chapters in the book where other issues are discussed further. We have looked at how the structures of governments are changing as they grapple with need for a coordinated approach to the environment but also with a need to decentralise decisions and management. Yencken (2002) advocates a whole-of government approach to environmental matters to counter some of the issues of coordination, overlap and underlap of responsibilities. Government bureaucracies use many different methods of environmental decision analysis and we have discussed two top-level approaches. Economic cost benefit analysis (CBA) can and should include environmental costs and benefits in addition to economic ones. Strategic environmental assessment (SEA) is a part of a top down approach to considering environmental aspects at a legislative/regulatory/ policy level leading to cumulative environmental impact assessment and to environmental impact assessment of individual projects.

Partnerships are being developed within countries and between countries to help address major environmental issues especially those issues that extend beyond local boundaries and beyond individual countries. Business organisations need to understand these, participate in them and be able to respond to them.

We have skimmed a very big subject but an important one. Business organisations have to deal with a variety of government departments on environmental matters and their ongoing operations are contingent on them obtaining approvals, licenses and reporting performance and compliance to government.

6.6 Further reading

Government Web sites are a good source of summary information on how governments are structured, how environmental matters are considered and what environmental legislation and departments exist. Some (viewed 3 March 2005) include:

USA government : http://www.firstgov.gov
USA environment protection agency : http://www.epa.gov
European Union : http://www.europa.eu.int
United Kingdom government : http://www.direct.gov.uk
United Kingdom major department : http://www.defra.gov.uk
French government : http://www.premier-ministre.gouv.fr/en/
Australian environmental agency : http://www.deh.gov.au

6.7 Questions

- For your country, identify the government departments with prime responsibility for environmental matters and try to find out how they undertake strategic environmental assessment.

7
Legislation and Institutions

Patricia Ryan

7.1 Introduction

This chapter discusses the legislative context in which environmental requirements and expectations impact upon the operations and strategic direction of business organisations. Legislation implements policy within institutional frameworks at government, community and corporate levels, with different operational and strategic implications for individual business organisations.

In particular, the changing nature and role of the legislative context, and changing inter-relationships among its institutional frameworks, affect how corporate decision-makers may influence environmental outcomes.

Organisational stances about environmental outcomes may partly be attributed to the extent of conformity between different institutions and between interpretations of regulation, self-regulation and ethics (Ryan, 2004a, pp. 91–4), as well as differing notions of sustainability and adaptive options for organisations (Ryan, 2003). Relevant institutions include governmental, legislative and judicial authorities, the mass media, non-government organisations and industry bodies, as well as internal norms and standards of conduct set by individual organisations (Ryan, 2004a, p. 86).

7.2 History of environmental legislation

Legislation with a primary focus on environmental issues largely dates from the 1970s (Fowler, 1984; Grinlinton, 1990), inspired by the landmark National Environmental Policy Act (NEPA) from the United States (US).

Regardless of whether national jurisdictions are codified, accustomed to the liberalism of Common Law traditions or subjected to despotism, environmental legislation has generally progressed from statutes dealing with specific concerns. In Australia, pre-20th century statutes were focussed on public lands, utilities and health. Early 20th century conservation of some natural resources was followed, mid-century, by town planning and limited development control over natural resources. By the 1970s, pollution control was an environmental, rather than health, issue. Broad environmental planning and assessment emerged in the 1980s.

Within the Australian federal structure, planning legislation achieved different styles at state and territory levels (Ryan, 1988), setting a pattern for later environmental legislation. Despite an end-of-century, federally co-operative approach centred on ecologically sustainable development (ESD) in response to international agreements (Bates, 1995, pp. 24–35), local constitutional and political factors prevented the national-level reach of federal legislation as in the US (Bates, 1995, p. 77).

Practical politics have also diluted once-strong legislation (Robinson, 1996) and resonate with 'smart state' images that promote 'constructive' industry relations (Grant and Papadakis, 2004) with financial incentives and market-based schemes. Related issues about community input and the capacity and reliability of information systems have had variable treatment in legislative provisions for appeals and judicial review of regulatory decisions, mandatory auditing, state-of-the-environment reporting and public-right-to-know.

7.3 Objectives of environmental legislation

Attempts to balance human activity with environmental concerns have occurred worldwide with development, not conservation, being the driving force. Electricity deregulation in the US and Great Britain illustrates how the deregulation lobby has ignored accumulating evidence of deregulatory failure in competitive terms, as much as from environmental and consumer perspectives (Palast, 2003). Certainly too, liberal development policies in developing nations have exacted a devastating social and environmental toll (Roy, 2002; Ryan and Wayuparb, 2004b).

Environmental legislation aims to preclude, or minimise, environmentally damaging choices, especially by government and industry, but also by consumers and households. Fears that coercive, conservation-oriented legislation will produce a worse-off society by limiting economic development and cultural, consumer and labour choice are offset by beliefs in the need for urgent, planetary-saving prescriptions. Just as legislation helped promote development, it ought to play a role in establishing an alternative environmental paradigm (Leane, 1998, pp. 28–9).

While legislation, like markets, is contorted by factions and free-riders (Krier, 1992), legislation is uniquely capable of an environmental orientation to tailor concepts, rights, responsibilities, processes, enforcement and dispute-resolution mechanisms, both across and beyond points of the market value-chain. Consequentially, some private property rights and discretions need to be restricted, redistributed or assigned to public institutions in the public interest.

Adequate delineation of this public interest is difficult and is a major source of friction regarding the acceptability and operation of environmental legislation, whose functions apply some pressures and relieve others. A public interest orientation may produce arbitrary and pragmatic boundaries in the absence of effective community input. The official status of legislation also means that the importance of legislative regimes can be over-stated, even misused, by those relying on it for legitimacy and can be downplayed by others.

7.4 Legislative impact on business organisations and directors

The role of legislation in contributing to environmental outcomes is not clear-cut. Legislation, as a politico-legal instrument, lacks system at the point of creation and operation. Its attitude-honing role comes largely from its words and formal legal and bureaucratic status. Even where corruption is not active, the wider the institutional context the more that discretionary processes of legislation are vulnerable to influential personal values.

Legislative scope

In Common Law systems, such as in Australia and former British colonies, case law is pivotal for interpreting and applying legislation to real people and events. Laws of evidence, procedure and statutory interpretation, as well as constitutional and administrative laws, act as litigation filters in a system that is neither concerned to seek a global reach nor adequate to appraise scientific, technological and managerial advancements.

Freedom from regulation, which is a purported attraction of globalisation, occurs because of: powerful lobbying; disagreement about the need for rules; and probative and enforcement difficulties, such as when multinationals have no property or real presence in the jurisdiction whose laws have been violated. Blind preferences for market conditions, however, ignore that legislation may assist both economic and environmental outcomes, by curtailing market abuses and conditioning the capacity for effective self-regulation.

Legislative awareness

In practice, interrelated environmental and legislative awareness among government and business decision-makers directly affects environmental performance regarding compliance with current legislation and anticipation of future legislation. Indirectly, legislative awareness affects organisational environmental practices that require environmental responsibility for resources, wastes, products and services, sites, marketing, and R&D. Legislative awareness, being connected with proactive environmental management, has also long been a major motivator of voluntary environmental audits (Coopers & Lybrand, 1991).

Such awareness by an organisation does not necessarily assure good alignment between environmental outcomes and organisational processes and structures. Legislation might still be ignored or merely prompt minimal compliance. Even highly coercive environmental legislation will not necessarily evoke a comprehensive management response.

Effectiveness

Legislation itself rarely resolves conflict, but relies on legal processes of determining particular winners and losers. 'Losers' may be alienated from the 'rules of the game' containing conflict resolution structures (Hall, 2000, p. 177). Alternative conflict management processes require trust and confidence in the reliability of persons, systems and principles (Giddens, 1990, p. 34), to enable collective action toward a common goal. Yet, most institutional arrangements exclude community-level resolution as a component of sustainability (Hall, 2000, pp. 178–9).

The ultimate test of effectiveness of environmental policy should be that legislation is not needed. Business, however, is thought to have largely been a follower, not leader, in environmental protection and to be influenced by how government behaves (Bowman and Davis, 1989; Cranston, 1984; Harrison and Antweiler, 2003; Kingham, 1995; Slavich, 1993; Vaughan et al, 1994). Training, education and a voluntary approach appear unsuccessful when isolated from interventionist approaches (Gunningham and Sinclair, 2004a), whereas successful intervention possibly requires hitherto unknown environmental commitment from the financial sector (Richardson, 2002) and judicious blending of positive and negative incentives (Gunningham and Sinclair, 2004b).

Directors

Focus on deterrence (Kane, 1991; c.f., Woodka, 1992) has led to penalties being imposed on directors and other corporate officers, irrespective of their personal culpability, unless mitigated by 'due diligence' (Howard, 2000). Actual penalties, however, need not match the popularity (Cohen, 1992; c.f., Howard, 2000) or intended seriousness (Hain and Cocklin, 2001) of prosecutions. In any event, penalties are possibly more symbolic in affirming community values, than in driving compliance (Farrier, 1992, pp. 86–8).

7.5 Public sector bodies: roles and business interface

Concern about weak or non-existent processes for enabling real community debate in development control systems is intensified where public sector bodies play a major discretionary role regarding activities by themselves or other public sector bodies. A related concern is where public sector regulatory bodies resort to bargains and compromises for securing compliance, instead of pursuing penalties against offenders, particularly those they consider 'reputable' (Farrier, 1992, pp. 88–98). Legislative mandate and design, resourcing, structure and formal checks-and-balances (Cranston, 1984, pp. 23–7; Gouldson and Murphy, 1998, pp. 48–52) might not curb extensive discretionary powers.

Public sector administrative roles rarely follow a single model (and include discouragement, encouragement, facilitation, regulation, prohibition and command), and business opportunities for interaction usually depend on whether public sector bodies directly set or impose standards, determine licences or permits, or exercise coercive roles independently of judicial proceedings or public inquiries. The nature of any business interface is affected by how public sector bodies strike a balance between being reactive and proactive, having regard to their historical practices, relationships and attitudes about their roles, including possible capture by the mindsets and cultures of the businesses they regulate (Cranston, 1984, pp. 27–33).

Co-regulatory approaches, relying on mutual openness and flexibility, may extend relationships between business organisations and public sector bodies, such as through consultative planning on future legislation, education/training, synergies with other government bodies and harmonisation between government program objectives and business profit and quality objectives (Ryan, 2003, p. 266). Additionally, government and business have to cooperate in demarcating, for instance, between self-interest and broader institutional interests and between neglect and inaction, in order to assist forward planning and not entrench unsustainable practices (Ryan, 2003, pp. 271–2).

7.6   Current legislation: objectives and issues

Vague or inconsistent environmental objectives in current legislation, combined with inadequate business agendas and inappropriate mechanisms (Gunningham and Sinclair, 2004b), can negate any prospects of 'perfect implementation' (Gouldson and Murphy, 1998, p. 13) of environmental policy.

Approvals and sanctions

The two most common mechanisms for achieving environmental policy (approvals and sanctions) also have significant drawbacks. Approvals for environmentally harmful activities sometimes promote a message of permissive legalisation or government complicity with business. Use of sanctions admits to failure of the other mechanisms and the traditional criminal law sanction model is more concerned with individual culpability than achievement of good environmental outcomes.

Yet legislative instruments need such backing in order to provide some certainty, particularly with respect to avoiding pollution, maintaining environmental capacity, achieving community expectations and establishing level playing fields. Legislation, therefore, increasingly employs tailor-made mechanisms such as:

- express objects and management principles;
- environmental impact assessment;
- open, or third party, standing for enforcement and appeals;
- specialist bodies and inquiry mechanisms;
- inducements such as performance bonds and tax concessions;
- education via labelling, registration and audit;
- special insurance funds and charges;
- due care and honesty requirements for managers and lenders; and
- corporate environmental reporting.

Current issues

Generally, however, legislative systems have failed to strike a balance between economic and environmental considerations or to optimise strategies for delivering environmental outcomes. Competing ideologies remain evident in inconsistent legislative definitions of 'environment', even in the same jurisdiction (Ryan, 2001, pp. 565–6).

Clarification and integration of the growing numbers of environmental policies and mechanisms, across the range of government, community and business organisations, has become a major legislative challenge around the world (e.g., Freestone, 1996; Farrier, 2002; Gouldson and Murphy, 1998, pp. 75, 110; Matson, 2001).

7.7 Future legislation: issues and directions

Global and local issues

The Action Plan of the 2002 World Summit on Sustainable Development lacked concrete strategies, despite some consensus about targets and the earlier prioritisation, in 2000, of 'Millenium Goals'. The Global Conscience Conference (2004) emphasised the need for sustainability outcomes grounded in the practice of community-wide responsibility and efficiency, plus binding international rules on the environment and conduct of business organisations, which are not dominated by economic objectives.

Localised legislation likewise lacks an underpinning of practice-oriented principles, such as the responsibility and efficiency notions of *natural* capital (van den Bergh, 1996) and *cultural* capital (Throsby, 2000, p. 11). *Cultural* capital advances sustainability principles by highlighting material and nonmaterial well-being, maintenance of cultural systems and recognition of interdependence (Throsby, 2000, pp. 13–4). *Natural* capital seeks a radical re-conceptualisation of resources (Lovins and Lovins, 2000) and promotes practices to increase productivity of natural resources, adopt biologically inspired production models, use solutions-based models and reinvest in natural capital (Lovins et al, 1999).

Future directions

Future legislative objectives might thus aim to achieve practical regard for broad sustainability principles in a holistic, systemic framework that combines ideas of natural capital, cultural capital and human-made capital (Berkes and Folke, 1994, p. 132). This framework requires credible sustainability standards capable of being implemented, monitored, evaluated and reported, as well as an adaptive organisational culture in public sector bodies and business organisations.

Future legislation might also seek to incorporate principles recognising that confrontation leads to little conservation progress (Hall, 2000, p. 174). This requires stakeholder linkages and relationships in formulating, developing, implementing and evaluating the conservation process.

The consequential idea of combining western paradigms of resource management with traditional land-use approaches has relevance for both developed and developing nations. Core elements (Lal and Young , 2000, p. 214), include a comprehensive environmental duty of care; ecosystem-based tenure systems; and institutional contexts, which define and link economic value systems within the constraints of ecological and cultural functions. In this way, renewed localism might offset the detrimental effects of globalisation.

In addition, a substantial ongoing legislative challenge is to reduce regulatory costs and stimulate the flow of positive environmental benefits into the enhancement of economic growth, employment, technology and industry (Johnson, 1991, p. 140), using vastly improved fact-finding, analytical and forecasting techniques.

7.8   Business operations and strategic direction

Many political, legal, financial and technological institutions influence corporate motives and strategy. Legislative potential for leveraging or influencing core organisational competencies towards better environmental outcomes depends upon how an organisation perceives legislation in the wider milieu of currencies, cultures, legal and political systems, resource availability, demographics, technological development, financial facilities and ethical constructs.

Legislation can affect a business organisation's current operations and strategic direction at three levels: *mandatory* (because business must comply with current legislation); *proactive* (in anticipation of future legislative changes); and *ethical* (for moral or ethical reasons).

Mandatory

At its basic level, environmental management involves keeping organisations out of legal trouble. Compliance systems focus on meeting legislative and licensing requirements and due diligence systems focus on providing managerial defence of criminal proceedings. Both kinds of systems should fit the nature, needs and risks of specific organisations and be fully implemented up, down and across an organisation, with internal and external communications playing a vital role (Schaffer, 1992; Slavich, 1993). Simplistic cost internalisation, to avoid liability, is insufficient to improve decision making.

Proactive

Although moving from compliance to proactive management may yield cost savings, improve image and efficiency, and enable an organisation to impose and share its concerns and learning across the organisation and market value chain (Slavich, 1993), decision making biases may work against the achievement of good environmental outcomes (Bazerman and Hoffman, 1999). Often, influential worldviews in an organisation will offload environmental issues for some other time, or to someone else. An optimistic antipodean view would believe that all environmental problems will be solved by experts, science, technology and wealth-creation (Mercer, 1991, pp. 309–10).

Ethical

Expressions of environmental concern need not be for moral or ethical reasons (Guerrette, 1986), although ethical reasons generally include legal compliance. Criteria for judging green business awards suggest ethical attributes of sustainability leadership that both reach beyond mere compliance and measurably reduce environmental impact (Ryan, 2003, pp. 267–71). In an optimistic scenario, business organisations and community bodies, such as NGOs, would influence each other, legislation, attitudes and behaviour of groups in market value chains and non-traditional alliances, under advanced institutional arrangements that emphasised principles of natural justice, ecological responsibility and the right to a reasonable standard of living (O'Riordan, 1995, p. 6).

Value-wash, however, in the growing industry of business models (such as corporate social responsibility, corporate citizenship, cause-related marketing, strategic philanthropy and enlightened self-interest), purporting to incorporate social and environmental factors voluntarily, may disguise lesser motivations of self-interest or warm-glow (Buchanan and Ryan, 2002, pp. 9–11, 23–5). It does not matter what environmental and social programs an organisation institutes, if it, say, monopolises available local water to make its products.

Case studies, in any event, support the need for voluntary action to complement, not replace, mandatory regulation (Gouldson and Murphy, 1998, pp. 69, 143–56) and optimal mixes of policy instruments could obviate distinctions between mandatory and voluntary (Gouldson and Murphy, 1998, pp. 131–4).

7.9 Summary and implications for business organisations

This chapter has indicated how legislation and its institutional frameworks may assist decision-makers in business organisations to achieve better environmental outcomes, foremost where government and business programs have integrity and are aligned in their objects, design and implementation. Legislative pitfalls include loose or ambiguous objects; reluctance or inability to institute holistic regimes involving the broader community; and misaligned implementation strategies, such as untargeted penalties and charges.

Business strategic plans must incorporate management of the legislative context, including its pitfalls. This requires tailored, proactive integration of the legislative context into the management, business, policies, culture, people and other corporate systems of an organisation. This may, or may not, involve an ISO Environmental Management System accreditation (Ryan, 2003, p. 271; c.f. Gouldson and Murphy, 1998, pp. 96–102).

In less-than-perfect systems, the synergies of jointly managing the legislative context and striving for continuous environmental improvement do have the potential to return joint organisational and environmental rewards (Johannson, 1994; Slavich, 1993).

7.10 Questions

- Find international award schemes that recognise and reward organisations that go beyond minimum legal compliance, continuously improve their environmental performance and voluntarily report independently verified information about their performance. Suggested starting point: Ryan, 2003, p. 271.
- Devise the key components of a sustainability strategy that leverages the legislative context in an actual organisation.

7.11 Further reading

Text: Mottershead (2002). Internet: INECE (2004), IUCN (2004).

8
Economics and the Environment

James White[1]

8.1 Introduction

This chapter begins by looking at ways in which the environment is valued in economic terms. It discusses how these valuations can affect decision making by governments and how the results of these decisions can be implemented through legislation and other governmental policy instruments. We conclude with a discussion on why business organisations need to understand environmental economics and how the application of its principles can eventually affect business operations. The term *value* in this chapter refers to *economic value* normally measured in monetary terms.

8.2 The environment and economic value

All aspects of economic activity result in effects on the environment, and the nature and quality of the environment affects our ability to carry out different activities. The economy and the environment are inextricably linked e.g. the environment:

- provides the resources needed for all economic activity, including energy, water, raw materials, soils, gases, chemicals and foodstuffs;
- receives the waste from production processes and the waste left over from completed consumption; and
- provides the ecosystem services that support life for humans and all other life.

Environmental values are defined in different ways in the literature. One classification (Wills, 1997) divides environmental values into three types:

- *Direct use values* are derived from use as an input to production (e.g. natural resources) or from direct consumption as a good (recreational). These types

of uses are easily identified and, in some cases, can be measured in markets
e.g. timber, fishing, and recreation;

- *Indirect use values* are based on the services the environment supports in
  providing goods that have direct uses e.g. carbon fixation, soil formation,
  water purification, and amenity; and
- *Non-use values* reflect satisfaction or values that an individual gains from
  the knowledge that the environment exists, e.g. existence of a species, or
  from the environment being preserved for future generations. These are
  also referred to as existence values.

Environmental values can also be assessed based on the environment's cultural role. These values are most often associated with local cultures around
the world e.g. a landscape can be a crucial link to a culture's history or a
particular species can have an important symbolic meaning. Alternatively,
the state of the environment e.g. clean rivers to support fishing, can also be
essential to traditional cultural practices.

All the above values are expressed as the value of the environment
to people. However, the environment is often noted as having *intrinsic
value*. This value cannot be measured in economic terms. It is a
value that exists irrespective of the environment's role in providing satisfaction or services to individuals. Intrinsic value reflects the view that
the environment has value in and of itself, regardless of its value to
people.

8.3   Environmental values in GDP and in GPI

The calculation of a country's Gross Domestic Product (GDP) is based on
the trade of goods and services in markets. Non-marketed goods and
services do not appear in the GDP figures. Using Wills (1997) classification,
some direct use values are captured in GDP e.g. the value of timber traded,
water supplied or fish supplied. Other direct use values (e.g. recreation) may
not appear in the GDP figures.

There are two main reasons why GDP does not provide an accurate
picture from an environmental perspective:

- GDP measures flows, not stocks e.g. GDP records fish catch or grain
  harvest, but not the underlying fish stocks or soils. High levels of
  GDP can mask levels of resource extraction that are depleting stocks
  (natural capital) and are unsustainable (Public Accounts and Estimates
  Committee 1998; Onisto 1999); and
- GDP does not measure non-monetised aspects of activity e.g. polluting
  oil spills are recorded as increases in GDP because they generate clean-
  up activity, while the loss of unique species or harm to ecosystems are
  generally excluded from the GDP.

However, this does not mean that there is no indication of indirect use or non-use values in the national accounts. When GDP is broken down into sectoral contributions, some environment-related measures can be seen, such as environment protection expenditure (e.g. sewage treatment services), which indicates the value society places on public health and a measure of environmental quality. This is one method by which environmental values can be estimated in monetary terms. This and other types of 'environmental valuation' techniques are further described below.

The notion of a 'green' or environmentally corrected GDP indicator, sometimes called a Genuine Progress Indicator (GPI) has recently received attention internationally. The GPI adjusts the traditional GDP measure to account for environmental costs, the value of output not traded in the market, income inequality and social costs. As an example, estimates for Australia developed by the Australia Institute (Hamilton and Denniss, 2000) show the GPI rising with GDP until the late 1970s, then diverging and falling until the late 1980s when the GPI started to rise slightly.

8.4 Environmental economics

Economics is about the management or allocation of resources under conditions of scarcity. A primary focus of economics is the efficient allocation of scarce resources so that society's wellbeing is maximised. Environmental economics is a sub-branch of economics that, in part, recognises and attempts to value the environment in economic terms so that it is appropriately allocated. It applies two main principles to the environment:

- natural resources provided by the environment are scarce[2] and so have an opportunity cost;[3] and
- full and proper values must be attached to environmental assets to ensure efficient use of resources.

By applying these principles, environmental services are treated as 'commodities' or 'goods' that can be valued based on an individual's preferences.[4]

Valuing the environment in monetary terms

Prices are a major signal of relative scarcity in market economies. According to economic theory, price signals steer resources toward their highest-valued and most efficient end uses. The challenge to environmental economics is that environmental services are often unpriced or underpriced,[5] e.g. the environment provides: essential life-support services (air, water, food); energy and raw materials that make possible our living standards; and natural amenities that enhance the quality of our lives.

These services are enjoyed by society without necessarily being rationed. Individuals do not necessarily pay to enjoy them, or at least do not necessarily pay the full cost of harnessing them for human use. At the same time, these services arise from scarce resources that exist in limited quantities, especially if they are not well managed.

In order to manage the environment in a sustainable manner, the relative importance (or value) of environmental resources needs to be known by market and non-market sectors of the economy, in order to guide allocation decisions. Non-market sectors are government, household and voluntary sectors of the economy that make their decisions about resource allocation based on factors other than profit. By establishing values, expressed in monetary terms, an individual or society can make informed decisions regarding their use or allocation.

8.5 Environmental valuation

Environmental valuation is described by Turner et al (1994) as measuring *human preferences* for or against changes in the state of the environment, rather than valuing the environment per se, in an effort to maximise welfare. One obvious way in which an aspect of the environment can be valued is to assign property rights and create markets for environmental goods. Examples of this are: water rights, fisheries individual transferable quotas, or emissions trading schemes.

Non-market valuation techniques can be used also. Non-market monetary values can be estimated for the environment via three general categories of valuation techniques:[6] *market-based*; *revealed preference*; and *stated preference*. The first two techniques, market-based and revealed preference, are used to estimate environmental values from people's actual expenditure choices i.e. what they actually do, rather than what they say they would do.

Market based techniques include: defensive expenditures (money spent to prevent environmental harm); replacement or repair costs (money spent to restore environmental quality); and productivity change (change in output due to change in environmental quality, e.g. crop losses due to salinity). Market based techniques often look at how the availability of environmental services impacts productivity and household expenditures.

Revealed preference techniques are based on examining markets where environmental quality can be related to market prices. Techniques include: hedonic pricing (estimating the value of an environmental attribute by examining its effect on property prices); and travel costs (estimating the value of an environmental attribute based on how much people will pay to travel to it). These techniques are often employed to estimate environmental values associated with recreation and amenities e.g. clean beaches.

Stated preference techniques are survey-based methods. These techniques involve asking people how much income they are prepared to trade for a given change in environmental quality. The simplest and most widely known method is contingent valuation, which is a pure survey technique that asks people their willingness to pay for environmental improvements or their willingness to accept payment for environmental degradation. Other methods that have grown in use are choice modelling and conjoint analysis, which are based on survey participants ranking groups of options, rather than simply answering questions about their willingness to pay. These methods allow estimates to be developed based upon how respondents are willing to trade between environmental attributes.

Developing monetary estimates of environmental services or goods require primary research that is time and resource intensive. As a result, analysts often default to benefit transfer. Benefit transfer is the process of taking the results of an environmental valuation technique that has been used in one place and applying them to a different area. Although commonly used, it should be carefully assessed to ensure that the transfer is appropriate. Guidelines for benefit transfer can be found in several documents outlining good practices for conducting cost-benefit analysis (USA EPA, 2000).

8.6 Can economics value all environmental values?

The short answer is no, but this can result from either practical or ethical perspectives.

Where people believe there are *intrinsic values* of the environment separate to the preferences of the wider population, these cannot be valued because our environmental valuation techniques are based on either people's revealed or stated preferences.

Some people also consider the environment to have value greater than the sum of its parts. Turner et al (1994) describes a sense in which economic valuation of the environment will represent only a partial value. He suggests that healthy ecosystems must first exist before society can receive the economic value of services such as waste assimilation. The value of those services (i.e. their natural capital value), therefore, does not encompass the primary value of the overall system itself which holds everything together and makes those valued ecosystem services possible.

Environmental valuation can also be limited in a practical sense as well as an ethical sense. The valuation techniques may be limited by either: lack of data to support investigations of environmental values implied through markets or surrogate markets; or by survey participants not having a good understanding of environmental attributes about which they are being surveyed and the implications of possible changes.

8.7   Environmental values used in public decision making

To bring environmental values into decision making, people first need to be aware of those environmental values, and then to appreciate their relative weight in a decision. Environmental values can be brought into public decision making through:

- requiring explicit consideration of environmental values in legislation, statutory decision making, planning and regulatory processes;
- ensuring environmental valuations are carried out and incorporated into economic analyses which inform decision making;
- providing public education about the significance of environmental values to help set cultural and behavioural norms, ethics and codes of moral behaviour toward the environment;
- defining property rights and creating markets for environmental goods e.g. emissions trading or fisheries quota trading; and
- reforming pricing and taxation frameworks to better reflect environmental values.

One way environmental values can be explicitly incorporated into government decision making (Chapter 6) is by using cost-benefit analysis to assess the relative desirability of competing alternatives (e.g. alternative sources of energy) in terms of economic worth to society. This frequently involves valuing and including benefits and costs which are not exchanged in markets. While such valuations may not be perfect, they are an attempt to gain more objective and transparent valuations of environmental costs and benefits, which would otherwise be valued more subjectively by the decision-maker (often a politician or bureaucrat), or at worst be completely ignored (Sinden and Thampapillai, 1995).

Over the last decade, economic instruments (or market based instruments) have become more prominent in addressing poorly defined property rights. Tradeable permits (or emissions trading) represent a quantity-based means of rationing access through the creation of property rights (OECD, 2004a).

Tradeable permits can provide significant benefits in terms of economic efficiency and environmental effectiveness. They have been applied in a variety of different contexts, including controlling air pollution, water pollution, water scarcity and fisheries depletion. Emissions trading is increasingly being used to address the challenge of climate change e.g. a European Emissions Trading Scheme is to commence in January 2005 covering 12,000 companies and representing half of Europe's emissions of CO_2.

Environmentally related taxes introduce a price signal that helps ensure firms and households take into account the costs on the environment when they make production and consumption decisions. There is growing evidence on the effectiveness of these levies as a means to reduce damage to the environment (OECD, 2001).

There has been an increasing role for environmentally related taxes in OECD countries over the last decade (OECD, 2004b). An increasing number of countries, in particular in the European Union, have implemented comprehensive 'green tax reforms'. These reforms can be implemented by a series of complementary measures, such as restructuring existing taxes e.g. transport or energy, to reflect their polluting characteristics, or by the introduction of new taxes e.g. on water use and water pollution, waste, environmentally harmful chemicals. The reforms also involve removing or adjusting environmentally harmful fiscal provisions (Barde and Braathen, 2002). As well as reducing environmental damage, it is argued that by replacing distortionary taxes on production and labour with less-distorting environmental taxes, these tax reforms can result in an overall improvement in economic welfare through higher growth and job creation, thus yielding a so-called 'double dividend' (OECD, 2002). A range of views in the literature about the 'double dividend' concept, though, vary from strong support to outright dismissal.

8.8 Environmental economic value and business decision making

Environmental economic values are generally not used directly by business organisations in their decision making but environmental economic values are indirectly incorporated into their decision making when they comply with regulations and other instruments (including tradeable permits and environmentally related taxes) that reflect environmental values.

Other areas where environmental economic values are indirectly incorporated into business organisations' operations include: marketing green products (Chapters 9 and 13); developing environmental business strategies (Chapter 11); responding to ethical investment pressures (Chapter 9); implementing Environmental Management Systems (Chapter 12); mitigating environmental insurance costs; responding to stakeholder environmental concerns (Chapters 9 and 19); and public reporting of environmental performance (Chapter 16).

One could argue that many of these are the incorporation of community ethical values (and increasingly one hopes business ethical values) in relation to the environment, rather than the incorporation of quantitatively derived economic values.

8.9 Summary

This chapter has discussed ways in which the environment is valued in economic terms and how these values can affect decision making by governments with these decisions often being implemented through legislation and other governmental policy instruments. If business organisations wish to be at the forefront of the changing environmental market they need to understand environmental economics and how placing value on the environment may eventually lead to effects on business operations.

8.10 Further reading

Stretton (1999); Tietenberg (1992, 1998); Tisdell (2003).

Notes

1. Contributions to the paper were provided by Carolyn Davies and Naomi Standing, both previously with the Department of Environment and Conservation, NSW, Australia.
2. Within an economic framework, goods or services are generally treated as 'scarce' resources. This definition does not necessarily imply that the goods are difficult to find, rather all goods are available in constrained quantities. In other words, while some goods are 'free goods' (i.e. their supply is unconstrained – think of air or seawater), most goods do not exist in quantities that can satisfy every individual's wants or needs.
3. Nicholson (1995) defines opportunity cost as the true cost of any action 'measured by the value of the best alternative that must be foregone when the action is taken.' Opportunity cost is a fundamental concept in economics. It is grounded in the assumption that resources are scarce. Increased consumption of one good requires trade-offs in the consumption of other goods.
4. By treating environmental services as 'goods', the principle applies that more of a particular good is always preferred to less. This principle is an extension of assumptions that characterise rational behaviour. It is assumed that individuals, when rational, express preferences that maximise their welfare given constraints or circumstances.
5. Public goods are generally non-exclusive. Unlike private goods, an individual cannot be excluded by price alone from enjoying the benefits of public goods. For example, clean air is a public good. Once available, individuals cannot be prevented from enjoying clean air and its benefits, e.g., lower health risks, high visibility.
6. In addition to the listed methods, the Delphi technique is also an option. It produces estimates of environmental values though surveys of experts in the particular field, often iteratively. However, this method has limited defensibility. It is not grounded in empirical evidence and results cannot be replicated.

9

Consumers and Community

Michael Polonsky (9.2–9.4), Fiona Court (9.5–9.7), Rory Sullivan
and Craig Mackenzie (9.8)

9.1   Introduction

This chapter discusses some of the issues involved in addressing environmental matters raised by consumers and the community and how these
groups can influence the actions of business organisations. The topics of
green marketing (Section 9.2–9.4), community consultation (Section
9.5–9.7) and ethical investment (Section 9.8) are presented by experienced
authors. Later chapters (13, 19) discuss the responses needed from business
organisations.

9.2   Introducing green marketing

It is important to start out with a brief discussion of basic marketing to
establish a common point of reference for marketing issues, within an
environmental framework. The term marketing is used to describe activities that create value through voluntary exchange between parties (Kotler,
2003). This involves a business organisation designing an offering com-
prising a range of characteristics, referred to as the 4-P's (Product, Price,
Place, Promotion). These can be combined in many ways to meet the
needs of consumers or customers, which are evidenced by the diverse
range of products available within a product category.

The marketing definition applies to environmentally responsible products as well, but organisations can operationalise this in many ways. Do
they produce environmental and *traditional* products in the same category
(Anonymous, 1990)? Do they shift their entire emphasise to produce *green*
products?

The inclusion of environmental attributes into marketing activities has
been discussed in a number of academic works in journals (Menon and
Menon, 1997; Polonsky and Rosenberger III, 2001a), books focusing on
this issues (Ottman, 1998; Fuller, 1999), as well as in the popular press
(Anonymous, 1991). It would appear that organisations have taken a

diverse range of approaches to using green marketing activities (Crain, 2000, Polonsky and Rosenberger III, 2001a) and have undertaken these activities for various reasons (Menon and Menon, 1997). In some cases organisations have embraced environmental issues because they have recognised that they have a duty to behave responsibly (Dimitri and Greene, 2002; Drumwright, 1994). E.g., the Australian producer Blackmores has had a longstanding environmental emphasis because of an ideological or emotive slant (Polonsky and Rosenberger III, 2001a).

Some organisations seem to have undertaken limited integration of environmental issues into their activities simply to differentiate themselves from competitors, without necessarily making any substantive change in their environmental activities (Polonsky et al, 1997). The rational for such superficial changes is that organisations believe consumers seek to make their consumption more responsible and will therefore be attracted to *green* products (Ottman, 1998). In many cases consumers rely on organisations' marketing activities for product environmental information, simply because it is difficult for them to evaluate products' environmental characteristics (Latvala and Kola, 2000; Morris et al, 1995), which is referred to as the credence characteristics of the information.

Unfortunately, organisations that position themselves as *green*, even when their activities are not green, or are no greener that competitors, result in general consumer skepticism of green claims, which in turn may lead to difficulties for all organisations who seek to leverage their products' environmental attributes in the market place (Carlson et al, 1996; Crain, 2000; Davis, 1993b; Polonsky and Rosenberger III, 2001a). Making environmental claims without any substantive support has been termed 'greenwash' and is one of the reasons that governments around the world have sought to regulate green marketing activities (Kangun and Polonsky, 1995).

9.3 Pressures to be green

There is an increasing realisation that humankind needs to consider its impact on the natural environment (Fuller, 1999). With six billion people in the world there are increasing economic and environmental pressures to meet these needs. It is suggested that these pressures will in fact get worse, as those in developing countries raise their consumption expectations to have lifestyles similar to people in the developed world (Singer, 1992).

The environmental consequences of such development will possibly be catastrophic, e.g. what would be the impact of China having a car ownership level equal to the US? There would be a massive increase in the production of greenhouse gasses from the use of automobiles and from their manufacture. In addition there would need to be a massive increase in expenditure on infrastructure to handle these levels of automobiles. While these activities might bring about economic 'growth' and improvements to living standards, there would also be substantial cost on society, increased road deaths, degradation of the natural environment and a congestion of cities, all of which could reduce the 'quality' of life.

Whose responsibility is it to ensure that human impact on the environment is 'managed' in a way that optimises the outcomes for humanity and the natural environment? It might be suggested that governments have the responsibility to maintain and protect the interests of their citizens of today and the future (Starik, 1995a), but there are numerous examples of where governments seeking to maximise economic outcomes discount the environmental impact of their decisions (Cook, 2002).

There is the added complexity that one nation's economic (and social) decisions impact on other countries. Early environmental problems such as acid rain were found to be unrelated to production or consumption activities in the countries affected, but were the result of production in neighbouring countries (Alm, 2000). One country cannot deal with global environmental issues, but rather there needs to be global coordinated environmental action. Such coordinated activities have had varying degrees of success, e.g. a number of countries have still not signed off on the Kyoto treaty to reduce greenhouse gases, which affect all countries (Gardiner, 2004). Other less publicised environmental agreements are also under dispute, e.g. there is a growing pressure to lift the moratorium on the commercial hunting of whales (Harrop, 2003).

A macro perspective assumes that governments can in fact drive consumer and organisational behaviour. However, even when organisations are required to modify actions they have a variety of alternative courses of action. Within the automotive industry there are different approaches being adopted in regards to the greening of automobiles e.g. Toyota has taken the approach of integrating environmental issues into its overall organisational philosophy and into its decisions (Polonsky and Rosenberger III, 2001a). This proactive stance results in a number of benefits; including these organisations being more innovative and thus able to cope with changes in business conditions, improving the efficiency of the production process and the products performance, while at the same time serving the needs of a broader cross-section of consumers. In this way environmental issues are a core part of decisions and not an afterthought. However, other manufacturers do not see greening as such a core issue and only seek to comply with national standards. This reactive posture results in organisations possibly being less innovative and thus less competitive in the long run (Porter and van der Linde, 1995).

Governmental policies also frequently seek to change consumer behaviour, but such changes are rarely universally embraced by the masses. Simple activities such as requiring consumers to recycle plastic bottles and paper, often encounters opposition from consumers, even when the effort and costs are minimal. Consumers are reluctant to change behaviour, which might partly be the result of them not understanding the implications of these changes. E.g., it is more cost and environmentally efficient for consumers to participate in schemes whereby they share goods, such as cars, yet for the most part consumers are hesitant to participate, because of a perceived loss of convenience or because they perceive that there is value from owning goods (Byrne and Polonsky, 2001; Oskamp, 1995). Changing the role of ownership of goods and consumption to purely utilitarian (i.e. functional value) is something that is counter to consumers self-identity and organisations' objectives of building brand equity, whereby that equity is reflected by status.

As such the greening of marketing needs to take a holistic approach that considers all stakeholders in the exchange process. An integrated approach will require a broader way of dealing with environmental issues and will most likely require compromise by many. The critical issue is that there is some inclusion of environmental issues and these do not become subservient to other financial objectives. There may need to be a system wide shift in thinking for this to occur (Fuller, 1999; Kilbourne et al, 2002). At the very least there needs to be a rethinking of the role of the environment in the broader exchange (i.e. marketing) process.

9.4 Green consumers

While there is some general scientific agreement that the environmental impacts of production and consumption are important, there are still varying degrees of consumer concern in regards to this issue (Kilbourne and Beckman, 1998; Ottman, 1998).

Leveraging the environmental issue with consumers is complex, especially when there are many complex issues that might be considered. Some consumers may be more concerned with issues associated with the production of greenhouse gasses, others with bio-diversity issues and others with water related issues. How are organisations to communicate complex sets of information to consumers in a way that is meaningful to these consumers? The difficulty is making these messages relevant to each group (or segment), but this assumes that consumer segments can be identified and communicated with, and that they will understand the information they are being provided, as well as act on the information (Kilbourne, 1995).

Evaluating a complex set of product information simultaneously is a complex process and environmental information is not considered in isolation to other information, such as price, quality, branding etc (Newell et al, 1998). In some cases 'environmental goods' are seen as having intrinsic value, which consumers are willing to pay for. E.g., eco-tourism is frequently seen as a more specialised and higher quality experience (Font and Tribe, 2001) possibly justifying higher prices. In other cases, even proven economic value does not encourage consumer purchases (Meijnders et al, 2001), e.g. consumers do not desire to 'invest' in long life light bulbs that have a significantly higher return on investment, as consumers traditionally only consider the 'purchase price' of the bulb and ignore the operating costs. As mentioned earlier consumers generally do not want to give up ownership of a car, even though it can be demonstrated that using public transportation and/or other private transportation (such as shared cars) is less environmentally harmful and less expensive. Thus consumers value their convenience, flexibility and possibly the prestige more than any environmental value (Byrne and Polonsky, 2001).

Getting consumers to embrace green product attributes or even include the environmental cost of owning/operating products in their purchasing process is something that will be difficult to achieve and may require long-term multiparty communication – business organisations, governments and non-governmental bodies. Unfortunately, most organisations do not see social change as part of their role, unless they see some direct organisational benefit. Thus, the majority of organisations do not seek to make such sweeping shifts in consumer behaviour and it is therefore unclear how such changes in behaviour will be brought about. In this less than ideal situation, one would expect that the majority of organisations will target the issues and activities that are most salient to consumers (Polonsky and Rosenberger III, 2001a). Real changes in attitudes and behaviors might require system wide 'shocks', such as the oil crisis of the 1970s, which resulted in a concern with automobile fuel efficiency.

9.5 Community involvement

'No man is prejudiced in favour of a thing knowing it to be wrong. He is attached to it in the belief of its being right,' (Thomas Paine 1792). Paine (2004) could easily have been describing the difficult process of community and stakeholder consultation. It is this strong attachment humans have to their beliefs and values that makes the process of consultation ironic – it both restricts our ability to negotiate in rational debate, yet it is the reason behind the need for consultation and communication. Community consultation sits within a framework of policy development, democratic process and government and industry decision making. The development of the process of community involvement in decisions that may affect people results in our current understanding of a community's 'right' to have 'a say'.

Out of a decade of struggle for urban renewal and poverty relief, Arnstein (1969) developed a theory on the various levels of community consultation. She described it as a scale of legitimacy, and it was subsequently called the ladder of consultation:

- degree of citizen power – citizen control, delegated power, partnership;
- degree of tokenism – placation, consultation, informing; and
- non-participation – therapy, manipulation.

Table 9.1   Benefits and problems of consultation

A transparent and robust consultation might:	The consultation process may also result in:
• Identify the stakeholders	• Political paralysis
• Define the agenda	• Time and cost consumption
• Improve information	• The creation of a forum for unrealistic expectations or opposition
• Ensure views are exchanged	• The removal of representativeness
• Create better decisions and a legitimacy for decision making	• The spread of inaccurate information
• Result in compliance with implementation	• The silence of the majority and the vocal influence of a minority
• Avoid future challenges	

Source: Based upon the NSW Government (1998).

The International Association for Public Participation (IAP2) defines consultation as: '...any process that involves the public in problem solving or decision making and uses public input to make better decisions. While there is an element of dispute resolution in all public participation, the essence of public participation is to begin a participatory process before the dispute arises' (IAP2, 2000a).

Reasons for undertaking community and stakeholder consultation include: making better policy decisions; seeking political support; trying to reduce environmental (including social and economic) impacts; and reducing the level of subsequent challenge to a decision. Benefits and problems of involving the community in decision making are shown in Table 9.1

9.6   Social and environment groups and consultation

Environmental groups have a strong role to play in community and stakeholder consultation. The importance of community involvement in decisions that may impact them has long been a key platform for such groups. The issues are increased transparency in decision making, and to ensure that the environmental impacts of a proposal, project or strategy are minimised.

The issue of scientific integrity and opinion is a critical one. Access to scientific opinion has increased sharply with household availability of the Internet. Diversity of scientific opinion has also increased – unlimited web publishing of reports, papers and 'positions'. This includes both accredited, tested analysis as well as unsubstantiated or untested information. Science is now readily available to counter science. It is difficult for many people to

---

Case Study 1:   Establishing environmental benchmarks

The involvement of environmental groups in the development of the Sydney 2000 green Olympics bid was a good example. Greenpeace has been tracking the games for over a decade and in 1993 the NSW Government awarded Greenpeace Australia best design for the Athletes Village. Further, an Environment Committee was established that provided a partnering role for environmental representatives and the Olympic Organising Committee. The subsequent *Environment Guidelines for the Summer Olympic Games* were officially adopted. They were monitored by Green Games Watch 2000, a coalition of environmental groups, as a third tier of Audit (Nature Conservation Council, 2001).

---

understand they have to compare the assumptions made, or the scenario examined, with their base case, and it is all too easy to take up a conclusion and apply it to a completely different scenario.

In response, organisations must employ both 'good science' and, equally importantly, a high level of transparency. The latter demonstrates a willingness to have the science used for project development 'tested' and a willingness to share assumptions and ideas used. Environmental groups offer a high level of representativeness. They do not just stand for themselves but a philosophy, knowledge, ethos and amalgamation of ideas held by a larger, silent, group of stakeholders. Increasingly, government and business organisations are choosing to partner with social or environment groups. The notion is that these groups accompany decision makers through their assessment and development processes, and can then provide a public endorsement of the strategy or product. The methodology or the science behind the decision making has been reviewed by this independent group (with green credentials) each step of the way.

9.7   Consultation with the wider community

The best time to think about community and stakeholder involvement is at the beginning of concept development of a project (or program) when you will have the most project 'negotiables' and design decisions have not taken you down an irreversible path. It is at this point you can ask – who should be involved, who would *want* to be involved, how, and what are the project risks and benefits? Possible partnering initiatives can be considered at this point.

Nimbyism

A community reaction to the pace of social and technological change is apparent. Embracing change is not an easy matter when, as a society, we are struggling with the nature, equity, frequency and speed of change. Change makes us question our social values – *'how does this new situation*

Case Study 2:   Establishing social benchmarks

An example of consultation influencing company policy and product to meet social and environment outcomes is the Telstra Consumer Consultative Council, which constitutes a range of social and consumer advocacy groups: Federation of Ethnic Communities Councils Australia; Central Land Council (an Aboriginal community group); and Australian Pensioners and Superannuants Federation. This Council helped to develop the *Access for Everyone* program, which focusses on providing affordable communications and has a role in the subsequent monitoring of Telstra's performance. Consultations like these make Telstra's self performance evaluation that much more meaningful (Telstra 2004).

---

affect me, my neighbours, or my community?' It is easier for us to stick with those parts of our lives that are constant, be it a tax system or the street where we live.

The public rejection of strategies or projects led to project managers developing the acronym NIMBY, which stands for Not In My Back Yard. This term was applied to people that opposed a proposal that was located near them. This and several other inventive acronyms became a way for many project managers to avoid the issues at heart: has the project been properly explained; has it taken into account the interests of those that could be directly affected; was the process one in which people could participate and negotiate; was the best outcome achieved; and could there be project flaws?

Community consultation demands a shift in attention from substance to process. That is, instead of immediately focusing on a possible solution, the project manager is forced to evaluate: what is the need for the project? How do I quantify the outcomes we are seeking? What processes are needed to make a decision? How can I constructively involve stakeholders in such a decision?

The public interest

Projects that affect sections of the community subsequently call into question the notion of 'public interest'. Consultation processes and public debate often pit an affected sub-group against a wider community, e.g. an abattoir is planned. Such a facility is needed by the wider community; it provides employment, an income for our shareholders, and food for the table. Yet when the wind blows in a certain direction over a relatively small amount of time each year, odours reach a nearby neighbourhood including a high school. What is in the public's best interest? How do we define the public interest?

The International Association of Public Participation definition is '...any group or individual, organisation or political entity with a stake in the outcome of a decision. They are often refereed to as external stakeholders.

They may be, or perceive that they may be, affected directly or indirectly by the outcome of a decision' (IAP2, 2000a).

The timescale and resources required for public discussion

The timescale of planning should also be relevant to achieving the considered response of stakeholders and the broader community. Obtaining a good feeling of community opinion may take years compared to the relatively shorter timescale devoted to design, assessment and approval (whether internal or external).

The 'decide, announce, defend' system of project development rarely works, as the ownership of ideas is essential to effective consultation. However, instigating a process that results in the public ownership of good ideas has timetable considerations (Chapter 19).

9.8 Ethical investment

Traditionally, investors have not paid much attention to ethical and environmental issues in their investment processes. While it has always been possible to find a few fund managers with an understanding of the commercial relevance of environmental and ethical issues, the reality has been that, for most asset managers, matters of corporate responsibility have not traditionally enjoyed much of a profile. This has started to change with both the Australian (Deni Greene, 2002) and UK markets (Eurosif, 2003) showing significant growth in ethical investment in particular with fund managers harnessing their position as investors to directly influence corporate policy. This differs from the more traditional screening approach to ethical investment, where investors include or exclude stocks on the basis of compliance or non-compliance with specified ethical or other criteria (Eurosif, 2004).

While directors' remuneration and board structure and composition have long been recognised as issues for shareholder attention, corporate responsibility or environmental issues have had nothing like the same profile, despite evidence that social, ethical and environmental issues can have significant impacts on company performance e.g. tobacco and asbestosis litigation, environmental regulation, potential damage to reputation or brand, and company failure as a consequence of probity failings (cf Enron). It is not only financial relevance that has started to drive investor interest in these issues. Regulation has proved to be another key driver. In Australia, under the *Financial Services Reform Act* (from March 2002), product disclosure statements for products with an investment component must include disclosure of 'the extent to which labour standards or environmental, social or ethical considerations are taken into account in the selection, retention or realisation of the investment'.

This parallels the change in the *UK Pensions Act* in 2000, which requires occupational pension funds to state the extent to which they take account of social, ethical and environmental issues with regard to investment. This led a large number of UK pension funds to change their Statement of Investment Principles (SIPs) to express a positive intention to take due account of these issues. As pensions funds are the largest clients of most asset managers, this has also led to a commercial pressure on fund managers to ensure that they have at least a basic capability to address these issues. However, even though many pension funds have changed their SIP to include social and environmental issues most have been slow to translate these changes into action (Coles and Green, 2002). Many company pension funds, for example, have not done very much to ensure their revised SIPs are being fully implemented by their appointed fund managers. Nevertheless, the foundations are now in place in the UK for a fundamentally more productive investor role in corporate responsibility.

Disclosure guidelines

One important development has been the suggestion that shareholders should hold companies accountable for effective risk management with regard to corporate responsibility issues. For example, the recently released Association of British Insurers (ABI) Disclosure Guidelines on Social Responsibility note that 'Institutional shareholders are also anxious to avoid unnecessary prescription or the imposition of costly burdens, which can unnecessarily restrict the ability of companies to generate returns. Indeed, by focusing on the need to identify and manage risks to the short and long-term value of the business from social, environmental and ethical matters, the guidelines highlight an opportunity to enhance value through appropriate response to these risks' (Association of British Insurers, 2002). The Guidelines state that companies should state in their annual reports whether the Board has:

- taken regular account of the significance of social, environmental and ethical (SEE) matters to the business of the company;
- identified and assessed the significant risks to the company's short and long term value from SEE matters and the opportunities to enhance value from an appropriate response;
- received adequate information to make this assessment and that account is taken of SEE matters in the training of directors; and
- ensured that the company has effective systems for managing significant risks, which, where relevant, incorporate performance management systems and appropriate remuneration incentives.

This approach has been welcomed by many shareholders, and many companies have responded positively. The risk approach is a valuable step but does not establish substantive standards for investors to measure corporate responsibility. This deliberate choice of the ABI was so that institutional shareholders should not put themselves in the position of making moral judgements which they are not qualified to make. There is, however, an ongoing discussion about the need for standards for holding companies to account. When faced with conflicting demands from pressure groups, customers, suppliers, governments and shareholders, companies want to know: what standards should legitimately guide their behaviour; when is there a genuine obligation for them to go further in addressing a problem: and when is there no such obligation. This is also a question for the shareholders because it is they who approve the appointment of the board, whose duty it is to conduct the company's business in the shareholders interests. When a company has either a genuine obligation to go further in meeting an ethical challenge or is facing illegitimate demands from extremist pressure groups, it deserves the full support of the shareholders.

Investor activity has been hampered by the lack of clear standards shareholders should expect of their companies. The most obvious answer is that companies should obey the law. In countries such as Australia, the abundance of regulatory protection for employees, consumers, the environment and the public means that if companies obey the letter and the spirit of the law, they should simultaneously meet their central corporate responsibilities. The law is not a perfect guide to corporate responsibility, but it provides an authoritative background of standards, against which shareholders can evaluate corporate responsibility. Unfortunately, this solution is not always effective. Recently severe criticisms with regard to corporate responsibility have arisen over companies' activities in developing countries. Corruption, conflict, poverty, disease, abysmal labour conditions, human rights abuses, and environmental problems – combined with the fact that many of these countries lack the capacity to impose effective business regulation – make it harder to do business responsibly in developing countries.

Corporate responsibility standards

In the absence of regulation, should shareholders define their 'own' standards for companies? While shareholder activism on corporate responsibility is of growing importance, it is in everyone's interest that such activism is based on a coherent and legitimate set of standards. An increasing number of investors refer to the framework provided by international law e.g. the Universal Declaration of Human Rights (UN, 1948), the Rio Declaration (UN, 1992) and conventions on climate change and biological diversity. What makes the application of these principles to companies problematic is that international law has almost exclusively concerned itself with the responsibilities of governments. There has been progress and the following references illustrate important first steps in codifying the social and environmental expectations of companies: UNCHR (2002); OECD (2000); and ILO (1977).

Greater clarity about the standards of corporate responsibility is an essential complement to the risk approach of ABI because where corporate responsibility is concerned, there is a tight relationship between risks and standards. Companies face the greatest corporate responsibility-related risks when their behaviour is out of line with the standards of behaviour considered legitimate e.g. risks to reputation (breach of standards considered legitimate by customers or commercial stakeholders) and legal (breach of standards considered legitimate by courts).

Shareholders are increasingly scrutinising the corporate responsibility performance of companies in which they invest. Shareholders need to have confidence that these risks are being managed properly and they expect that the importance of these issues is recognised by companies, that appropriate systems and processes are in place to manage these issues and that senior management takes responsibility for managing these issues. More work is needed here, and this is perhaps where shareholder engagement with companies should focus. The more we can move to greater clarity and consensus between business, government, investors and other stakeholders (e.g. NGOs, trade unions) about standards, the more surefooted companies will be at managing matters of corporate responsibility and its associated risks.

9.9   Summary

In this chapter the four authors have discussed three diverse ways in which consumers and community may demonstrate their environmental attitudes (positive and negative) towards the actions of business organisations: buying or not buying green products; supporting or opposing projects and operational practices; and investing or not investing with socially responsible organisations.

10
Business Management and the Environment
Robert Staib

10.1 Introduction

This chapter completes the first Part of the book. Chapters 2 and 9 have described some of the external environmental issues and influences that can affect the environmental performance of an organisation. Figure 1.1 encapsulated these externalities under the broad headings of Environment (physical and social) and the External Context (organisations and people). These are summarised in Table 10.1. With this background the chapter discusses some of the ways business has and is responding to environmental issues. Responses at the individual organisation level are discussed in later chapters of the book.

We start with a short history of corporate environmentalism to set the scene and then list some of the external groups that can influence an organisation's environmental management. Two examples are discussed:

Table 10.1   Environment aspects external to an organisation

Ch. Environmental aspect	Grouping (Figure 1.1)	Generic Description
ENVIRONMENT		
2 Environmental trends and impacts	*Physical*	The natural environment
3 Attitudes and behaviour	Social	Peoples' perception
EXTERNAL CONTEXT		
5 Politics and the environment	Organisations	The key government controlled organisations
6 Govt. environmental decisions	"	
7 Legislation and institutions	"	
8 Economics and the environment	"	
4 Ethics and belief systems	People	The general community
9 Consumers and community	"	

industry associations which articulate the environmental threats and opportunities for their member organisations; and insurers which cover environmental risks by a combination of insurance and risk management.

We discuss how many tradition management processes and systems are being changed to incorporate environmental aspects. This leads to a discussion on sustainability and how it might be applied at the organisational level rather than at the usual level of society as a whole.

10.2 Stages of corporate environmentalism

The history of corporate environmentalism is relatively short with most action occurring in industrialised society after the Second World War (Chapters 2 and 3). Authors like to categorise and slot past events into stages or phases and forecast the future. Environmental authors are no exception. Frankel (1998, p. 37) describes 4 eras of corporate environmentalism from 1962 to 1998, each loosely initiated or marked by salient or catastrophic environmental events e.g. the Bhopal chemical spill in India. Table 10.2 shows Frankel's eras and his assessment of their effect on business.

Hoffman (2001, p. 150) categorises 4 stages from 1960 to 1993 through which American organisations have been evolving and responding differently to environmental pressures. Earlier, Hoffman (1999) had linked the stages of corporate environmentalism with events similar to Frankel. Table 10.3 does not include all of the characteristics of Hoffman's categories but summarises an aspect of organisations that has changed in response to a changing environmental milieu: how environmental responsibilities have been progressively spreading throughout all levels and functions of organisations.

Tables 10.2 and 10.3 are American examples, are approximate and a bit arbitrary but do serve to illustrate the changing nature of the business response to changing environmental and community pressures. They should also be considered against a changing background of environmental degradation (Chapter 2) and changing community attitudes and behaviour (Chapter 3). Both Hoffman and Frankel believe that change will continue, must continue, if we are to maintain the natural environment of our planet.

Hawken et al (2000) foretells of the 'Next Industrial Revolution' which will be based on the highly efficient use, by business organisations, of the *Natural Capital* of the earth as distinct from the financial capital of capitalist society. Welford (2000, 1998, 1997) in his three linked books looks forward to better corporate environmental management leading organisations to develop environmental *systems and strategies*, change environmental *culture and organisation structure* and to achieve *sustainable development*. Dunphy et al (2003) guides us through a *transformational path* towards organisational change for corporate sustainability using a business strategic planning and management approach. These authors provide their vision for a better environmental future and describe how business needs to change to achieve it.

Table 10.2 Corporate environmentalism: eras

Era name	Era Start	Salient events	Description	Effect on Business
First	1962	Publication of Silent Spring (Carson, 1963)	Extensive environmental damage from pollution by chemicals	Compliance with laws and regulations
Second	1984	Union Carbide plant in Bhopal India	Accidental release of 57,000 litres of methyl isocyanate into environment	Corporate environmental practices change, more public accountability
	1989	Super tanker Exxon Valdez	Tanker ruptured and spilled 41.6 million litres of crude oil into Prince William Sound	Beyond compliance, aim to exceed emission levels
Third	1992	Eco-efficiency	Less materials and energy in, less waste out	Win-win, with less pollution, less resources and cost
	1992	Rio de Janeiro Earth Summit, Book: Changing Course, (Schmidheiny, 1992)	Focus on sustainable development	Sets course for business but reduced sustainable development to the lesser eco-efficiency
Fourth	2004 plus	Population growth, resource use and environmental degradation	Achievement of real sustainable development	Towards zero waste, whole system thinking, looking outward, real sustainable development.

Source: Frankel (1998, pp. 37–49).

Table 10.3 Corporate environmentalism: periods

Time	Type of environmentalism	Organisational structure and culture		Salient events
		Characteristic	Environmental responsibilities	
1960–1970	Industrial	Problem solving	Ancillary to business, primarily an operating function	1962 Silent Spring 1963 Mississippi fish kills
1970–1982	Regulatory	Technical compliance	Environment as a separate department but focussed on legal requirements	1970 Earth Day 1970 USA EPA formed
1982–1988	Social responsibility	Managerial compliance	Managerial structures changed and responsibilities spread throughout the organisation	1983 USA EPA Administrator dismissed
1988–1993	Strategic	Proactive management	Environmental department has greater power and environment integrated into processes and products	1984 Bhopal disaster 1985 Ozone hole 1987 Toxic Release Inventory 1987 Montreal Protocol 1987 Exxon Valdez
1993–2001	Sustainable development	Social equity	Corporate sustainability objectives and indicators set for social, financial, environmental impacts of the organisation	

Source: Hoffman (2001, p. 150, pp. 219–224; 1999).

10.3 External influencing groups

Before we discuss business sustainability further, we will look briefly at some of the groups external to an organisation that can influence the way in which an organisation does business, responds to environmental issues and plans for its future (Refer to Figure 10.1). The top row (govern-

Figure 10.1   External influencing groups
Source: Based on Hoffman (2001, pp. 202–206); Behrendt et al (1997).

ment, government organisations, and Standards organisations) impose environmental requirements. The next group (environmental or green groups, consultants [environmental, legal], financial institutions, shareholders, investors, insurers) are starting to make their environmental requirements and advice felt. The next group (competitors, community and the supply chain: suppliers and customers) have a more direct and immediate effect on an organisation's environmental direction and finally the last group (industry think-tanks, universities, authors and publishers, industry and trade associations) provide some of the knowledge and information that can assist an organisation to respond. Many of these groups are discussed in more detail in later sections of the book but it is worth discussing here as an example the influence of two of these groups.

Industry associations

Industry associations, business peak groups and trade associations are starting to extend their services to members to cover environmental issues: sometimes articulating the threats and issues for their member organisation; sometimes identifying opportunities.

At the international level, the World Business Council for Sustainable Development (WBCSD) was formed to plan an approach to the United Nations Conference on Environment and Development in Rio de Janeiro in 1992. It is now a coalition of 170 large international companies united by a shared commitment to sustainable development and believes that business is good for sustainable development and that sustainable development is good for business. Its mission is to provide business leadership as a catalyst for change toward sustainable development, and to promote the role of eco-efficiency, innovation and corporate social responsibility. Its objectives and strategic directions include: *business leadership* – to be the leading business advocate on issues connected with sustainable development; *policy development* – to participate in policy development in order to create a framework that allows business to contribute effectively to sustainable development; *best practice* – to demonstrate business progress in environmental and resource management and corporate social responsibility and to share leading-edge practices among its members; and *global outreach* – to contribute to a sustainable future for developing nations and nations in transition. It embraces three pillars: economic growth, ecological balance and social progress (World Business Council for Sustainable Development, 2004).

At the national level e.g. in Australia, there are associations like the Business Council of Australia and the Australian Industry Group. The Business Council of Australia is an association of Chief Executive Officers of private companies and is a WBCSD partner though environmental sustainability is only a part of its role in promoting a financially sustainable business. The Australian Industry Group is Australia's leading industry organisation representing 10,000 member companies in manufacturing, construction, automotive, telecommunications, IT, transport, labour hire and other industries. Its role is to assist Australian industry to become more competitive at the domestic and international level. As well it aims to keep its members up to date on environmental and energy issues that impact directly on their businesses and to help put in place environmentally friendly and energy efficient practices with a positive impact on the bottom line. It does not believe that in order to become more environmentally efficient companies must sacrifice their drive for profit (Australian Industry Group, 2004).

Insurers: risk and insurance

Increasingly organisations are being held liable for the environmental consequences of their actions through more stringent legislation. Tables 10.2 and 10.3 highlight some of the major accidents in the last 40 years with severe environmental consequences that have influenced the way organisations manage potential environmental risks. The environmental consequences could be sudden as in the case of the Bhopal incident or gradual as in the gradual build up of agricultural chemicals in the environment (Carson, 1963), where the environmental consequences took many years to manifest themselves.

As well as this sudden/gradual distinction, the insurance industry identifies two levels of environmental damage: primary damage – where there is ecological damage to air, water, soil, flora and fauna with no clear property rights and secondary damage – where there is personal and property damage (Swiss Re, 2003). This latter responsibility includes the environmental liability for managing pollutants from an organisation's own sites – both remediation of historic pollution and new pollution. This means that an organisation can also be held responsible for the pollution of its site even if it was caused by a previous owner especially where the previous owner can not be found or the previous source of pollution identified (Swiss Re, 1999).

The cost of insurance to cover the environmental consequences of organisations' actions on land can be high. Significant costs can also be incurred if an event occurs that is not covered by an organisation's existing policies e.g. pollution caused by previous occupiers of the land owned by the organisation manifesting itself or the organisation being found to be subsequently responsible for previous ecological damage. Rather than wait for an event to occur organisations can adopt a environmental risk management approach. This approach (sometimes called loss prevention) is an integrated approach that involves; assessment and control of risk factors; insurance to cover specific risks; and allocation of organisations reserves to cover other risks (self insurance) (Broderick et al, 2000; Swiss Re, 1999). Chapter 21 discusses environmental risk assessment and management.

Notwithstanding the enthusiastic embracing the combination of risk management and insurance by some of the authors quoted above, Minoli and Bell (2003) says that insurance (which includes an initial underwriting assessment of risk) in the United Kingdom context is a weak financial instrument to prevent pollution related losses. Schwarze (2001) though believes that the German Environmental Liability Act (which enforces strict liability for environmental damages) used in conjunction with appropriate environmental insurance and environmental risk assessment and management is an effective method of controlling environmental accidents and costs.

10.4   Business management processes and systems

Organisations both public and private need management processes and systems to function and to govern themselves. There are many different sorts of management systems within organisations either stand-alone systems (e.g. a project management system to manage cost and time aspects of projects) or system that link and integrate existing ones (e.g. quality assurance systems).

Environmental management systems

Systems have been developed to specifically address environmental issues within organisations. After initial adhoc approaches various structured and standard processes and systems started to appear. British Standard BS 7750 was one of the first to appear. It became the basis for the ISO 14000 standard that is being adopted world wide. ISO 14000 is a series of codes covering environmental management systems, auditing, environmental labels and declarations, environmental performance evaluation, life cycle assessment and environmental risk management.

Other environmental systems include the European Union standard Eco-Management and Audit Scheme (EMAS) for companies performing industrial activities with only discrete sites being able to be registered. Various industry schemes and systems have been established e.g. Responsible CARE is an environmental management process used by the chemical industry (Welford, 1998). Relevant corporate processes are discussed in Chapters 11 to 16 with Chapter 12 discussing environmental management systems.

Strategic environmental planning and management

Many organisations are now integrating their environmental management processes management into normal business processes. This can involve incorporation of environmental aspects into strategic planning and management, corporate social responsibility strategies, value chain analysis, customer satisfaction and marketing plans, organisational design, accounting and financial management. Strategic aspects are discussed in Chapter 11.

10.5 Industry sustainability

We have used the term sustainability rather loosely in this chapter. In recent years a number of industry wide approaches have emerged e.g. ecologically sustainable development (ESD), sustainable development, corporate social responsibility (CSR), eco-efficiency, industrial ecology and others. Starting from the concept of ecologically sustainable development (ESD) we have moved towards these other possibly lesser environmental concepts of sustainable development, sustainability and corporate social responsibility with its triptych of scenarios: social, economic and environmental. Some authors question the concept of sustainable development being a way business organisations can justify business as usual with some adjustments for the environment. See Section 5.5 for definitions of sustainability.

In the author's experience, the words 'sustainable development' commonly refer to development of better environmental practices and outcomes but rarely to real ecologically sustainable development. The indicators of

Chapter 1 do not demonstrate great steps towards ESD. Lamming et al (1999) suggests that while business organisations can contribute substantially to global sustainable development, they cannot individually achieve sustainable development because many environmental aspects are outside their control. Therefore setting sustainable development as an achievable organisational strategic goal should be carefully considered.

Other authors are saying that to save the environment we do need to move towards real sustainable development and that there are welcoming signs from some leading organisations that much can be and is being achieved (Frankel, 1998, Dunphy et al, 2003, Hawken et al, 2000). Hawken et al (2000, p. 11) uses the words of the Factor Ten Club in describing how within one generation we can achieve a ten-fold increase in efficiency in the way we use energy, natural resources and other materials. He quotes many examples of companies leading the way with substantial productivity gains. This is fine at the individual company level but how does this survive if one looks at Ehrlich et al's thesis that environmental impact increases in proportion to the product of population growth (with its con-comitant growth in demand) and increasingly material acquisitiveness (with more energy, material use and production of waste per person) (Ehrlich et al, 1977, p. 720). The quandary is that if an organisation can achieve a 50% reduction in material and energy use in its products but increase its sales by 50% the environment may be no better off, though it would be certainly better off than if the company achieved no productivity gain.

Ecological footprint

Much is being written on how to measure sustainable development at the national or international level through the use of indicators of the health of our ecosystems and our use of resources e.g. Environmental Sustain-ability Index (Esty and Samuel-Johnson, 2001) and Ecological Footprint (Wackernagel et al, circa 1997). The Ecological footprint is the biologically-productive area required to continuously provide resource and energy supplies and absorb wastes of a particular population given prevailing technology. Wackernagel's team (circa 1997) calculated the ecological footprints, both the impact footprints and the productive areas. The impact footprints ranged from USA with 10 ha/person to Bangladesh at 0.5 ha/person. They also calculated that the available productive land in the world at approximately 1.7 hectares per person in 1997 and the average world ecological footprint at 2.1 hectares per person meaning we are already in deficit. Table 10.4 shows a selection of their figures and includes the ecological deficit and the total country footprints.

Is it possible to apply this ecological footprint to the individual organi-sational level? It is theoretically possible to calculate the footprint of an

Table 10.4   Ecological footprints

Country	Impact footprint	Available capacity	Ecological Deficit	Population $\times 10^6$	Country footprint $km^2 \times 10^6$
	hectares per person				
USA	10.3	6.7	–3.6	268	27.6
Australia	9.0	14.0	5.0	19	1.7
United Kingdom	5.2	1.7	–3.5	58	3.1
France	4.1	4.2	0.1	58	2.4
Costa Rica	2.5	2.5	0.0	4	0.1
Egypt	1.2	0.2	–1.0	65	0.8
China	1.2	0.8	–0.4	1247	15.0
Bangladesh	0.5	0.3	–0.2	126	0.6

Source: Wackernagel et al (circa 1997).

organisation and its products and services though many assumptions are necessary in calculating the ecological footprint. The ecological footprint concept does not adequately address a number of issues including different risk levels, time aspects, use of natural capital and differences between non-renewable and renewable resources. It also requires gathering and maintaining much data. It may be useful for individual organisations as an indicator for comparison and trending analysis (Holland, 2003). Two water supply authorities have done so. Sydney Water in Australia at 0.017 ha/person (Sydney Water Corporation, 2002) and Anglian Water in the United Kingdom at 0.01 ha/person (Holland, 2003), though a direct comparison should only be done after reviewing the assumptions used in each calculation.

Despite much literature on the concepts of sustainable development and its measurement, the World Business Council for Sustainable Development believes that at this stage there is no 'proven recipe to achieve sustainable development' (Fussler, 2004).

Environmental targets

Notwithstanding that there are no accurate measures of sustainable development or no proven recipe, there are many different approaches to achieving substantial improvements in environmental performance of business organisations. At the basic level organisations should replace their qualitative goals like 'minimise impact or reduce waste' with quantitative goals like 'reduce waste by 90% in 4 years', and then move to more rigorous goals like 'zero discharge, no net energy use or use only renewable energy'. At the inspirational level they can adopt the approach of doing more with significantly less by making substantial increases in productivity in the use of natural resources, energy, human resources. Some of the authors quoted in this chapter provide this inspiration (Dunphy et al, 2003; Hawken et al, 2000; Frankel, 1998).

10.6 Summary

In this chapter we have discussed some of the ways business is responding to environmental issues. We started with a short history of corporate environmentalism to set the scene and then listed some of the external groups that influence an organisation's environmental management and as an example discussed how industry associations are articulating the environmental threats and opportunities for their member organisations and how insurers are seeking to cover environmental risks.

We saw that the concept of sustainable development is becoming a significant part of business life but at this stage, it is difficult to measure and to achieve. Notwithstanding, there is much business can do to significantly improve its environmental performance and to significantly reduce its impact on the natural environment. An organisation needs to start at the basic level of compliance with legislation and proceed progressively to more strategic approaches. The remaining chapters in the book discuss some of the management approaches needed including systems and tools that can be used.

10.7 Questions

Review the references quoted in the Case Study box. HP is embracing sustainability by changing its management approach and corporate goals and by seeking to make products durable, have longer life, be upgradeable and be recyclable. It is emphasising dematerialisation but relies on increased sales to make the company financially sustainable.

- Is the overall impact on the environment e.g. total volume of resources required (materials and energy in manufacture, use and reuse) to be reduced and limited to what the natural environment can deliver or is the reduction focussed more on limiting the resources per product?
- Does its approach create more consumer needs, more consumer demands and consequently more demands on the natural environment?
- Is HP's approach ecologically sustainability development (when viewed from the natural environment's point of view) or is it a sustainability with compromise between social, financial and environmental?

Case study: sustainability at Hewlett Packard

HP, a technology-based company, believes sustainability is a strategy that is imperative for business. HP's environmental response has evolved from *pollution control and prevention* through *product stewardship* and is heading towards *sustainability*. HP has established two goals: improve HP's environmental performance in today's products and processes; and invent tomorrows' sustainable business. These goals are supported by strategic initiatives: end-of-life product solutions; energy efficiency; dematerialisation; and the development of measures to assess progress towards its environmental goals. It is an environmental strategy that HP believes will increase customer demand (for environmentally sound products) and improve profitability. It is seeking to invent new solutions to address societies' needs with services rather than products i.e. a dematerialisation of its output. The environmental strategy integrates with a business strategy directed towards: meeting customer and market expectations; improving market access; increasing cost savings; creating market opportunities; enhancing brand image; levering competitive advantage; and increasing shareholder value (Preston, 2001).

Table 10.5 shows some of the environmental indicators measured by HP. Oekom Research AG, an independent rating agency, says that toxic materials in computers is a major environmental problem but says HP's performance is one of the best (Hewlett Packard, 2003).

Table 10.5 Hewlett-Packard environmental reporting

Measure	Unit	2000	2002
Revenue (less Compaq)	US$ $\times 10^9$	48.9	56.6
Greenhouse Gas release	tonne $\times 10^3$	233	241
Perfluocarbon release, Base Index 1 in 1995	index	1.74	1.97
Electricity use	GWh	1,714	1,709
Hazardous waste production	tonne	6,860	6,440
Other waste production	tonne $\times 10^3$	116	106
Toxic Release Inventory substances	tonne	1,130	868
Energy efficiency increase per unit, high end servers. Base Index 1 in 1998	index	0.9	1.9
Material efficiency increase per unit, non-stop servers. Base Index 1 in 1989	index	72	82

Part II
Corporate Processes and Systems

11
Corporate Environmental Strategy

Robert Staib

11.1 Introduction

This chapter presents some traditional approaches used for corporate strategic planning and management and discusses some ways being used to integrate environment issues into organisational strategy. We describe a strategic management process model and discuss aspects of this model in more detail (strategic analysis, direction, evaluation and choice) and leave the reader to investigate the other strategic processes of implementation and reanalysis. Under the evaluation aspect, two approaches to the measurement of strategic environmental performance are discussed: Dunphy et al's (2003) phase model *Organisational Change for Corporate Sustainability* and Kaplan and Norton's model the *Balanced Scorecard* (Kaplan and Norton, 1992).

11.2 Corporate strategic management

Business organisations will continue in the traditional way to:

- manage resources, manufacture goods, produce goods and services, provide employment;
- use environmental resources: energy, earth materials, plant and animal materials; and
- create pollution and generate waste;

and increasingly in a more environmentally responsible way:

- minimise/prevent pollution and avoid/reuse/recycle wastes;
- provide environmental goods and services;
- meet/create consumer demand for green goods and services and develop new green products; and
- reduce consumption of environmental resources.

Subject to the excesses of some business practices, most western countries accept that it is preferable that business organisations stay in business. Continuing to make a profit and continuing to grow appear to be the most accepted measures that will guarantee that this happens at least in the medium term. The approach to business strategy described in this chapter flows from these fundamental theses.

Approaches to strategic management and methodologies used by organisations are numerous and vary considerably with many organisations now integrating their environmental management processes into normal business processes. This integration should start at the strategic level. This can be done progressively (an incremental approach) or in a bolder way by placing environment at the centre of strategic planning and management (transformational approach) (Dunphy et al, 2003). The strategic paradigm of sustainable development that supports the transformational approach is becoming a rallying point for business organisations but the theory appears to be ahead of practice at the moment (Section 10.5).

Corporate strategic planning and management can be seen in terms of the Deming cycle (Figure 12.1) but on a grander scale and over a longer time frame. Table 11.1 summarises the processes followed in traditional

Table 11.1 Strategic management processes

STARTING POINT	VISION : our ideal future
------>⃓	MISSION : what we do
	VALUES : how we do things

Strategic process	Strategic actions
Analysis	• External environment • Internal skills and resources • Stakeholders needs and expectations
Direction	• Formulate objectives and targets • Identify performance measures
Choice	• Generate options • Choose preferred strategy
Implement	• Develop appropriate systems • Acquire and utilize skills and resources • Develop organisational structure • Manage the culture
Evaluation & control	• Measure performance • Take corrective action
Reanalysis	• Return to top of cyclical process

Source: Based upon Viljoen and Dann (2003), p. 37.

strategic planning and management (Viljoen and Dann, 2003). In the following sections we discuss the strategic processes: *analysis, direction, choice* and *evaluation* (e.g. measurement of performance) and leave the reader to investigate the other strategic processes of implementation and reanalysis.

11.3 Strategic analysis

Strategic management is often portrayed as a planning driven linear process with a recurring feedback loop (Column 1 in Table 11.1). In many situations it can be an iterative process with many stops and starts, with impetus coming from the external environment, being driven from the top of an organisation or pushed up from below. Subject to these limitations, it is worth considering some of the techniques (or processes) associated with this linear model that are used by organisations during *strategic analysis*. We will discuss a sample with the objective of considering how the environment might be brought into these analyses. A word of caution when discussing strategy and the environment: many authors on strategy use the word environment in a generic sense to mean those areas external to the business not necessarily the natural environment which is often treated as the regulatory environment (Johnson and Scholes, 2002, p. 97).

The time frame for many strategic environmental issues (i.e. the time taken for problems to manifest themselves and the time taken for their solution) is likely to be longer than that for business strategic issues. It is suggested that environmental strategic planning and management should look at least one generation ahead. The size of the societal and organisational change necessary to address and solve major environmental problems is large and often these large societal changes are generational i.e. occurring over 20 to 25 years (Hoffman, 2001, p. 139; Staib, 1997, p. 168).

Broad analysis

The *SWOT* analysis (identification of an organisation's strengths, weaknesses, opportunities and threats) is a common technique used for strategic analysis. It is fairly easy to use and may result in list of approximately 10 items against each heading. Some authors suggest that it is not that rigorous (Hill and Westbrook, 1997; Grundy, 2003, p. 10) and should not be the sole technique for strategic analysis. Notwithstanding, it can be a useful approach to identify some of the key issues that are likely to impinge upon an organisation's strategic direction (Johnson and Scholes, 2002, p. 183). It is easily adaptable to include strategic environmental issues especially if an organisation's compliance and environmental managers are involved as an integral part of the analysis.

The business external environment

Scanning of the business external environment is an important part of strategic analysis and can be done at the general level of political, economic, technological and social (*PEST*) (Grundy, 2003, p. 10). Here again environmental issues impinge upon all of these categories. Figure 10.1 shows some of the external organisations and stakeholders that can exert environmental pressure or influence which in turn can affect the strategic performance of an organisation.

Competitive forces

Porter's five forces is a useful model to explore changes in structure and dynamics of the industry (or business environment) in which an organisation operates. At its centre is an assessment of an industry's competitors (i.e. competition or rivalry among existing firms in that industry). Impinging upon and influencing an organisation's position to varying degrees are the: bargaining power of suppliers; threat of substitute products or services; bargaining power of buyers; potential entrants and entry barriers; and industry competitors i.e. the degree of rivalry (Porter, 1979; Viljoen and Dann, 2003, p. 133; Hubbard, 2000, p. 33).

Writing 16 years later on the effects of a growing amount of environmental legislation on organisations, Porter says that despite a belief that there needs to be a trade off between economy and environment there are significant process and product benefits in adopting an environmental approach to developing an organisation's strategy (Porter and van der Linde, 1995). He sees that linking an organisation's business strategy to the environment can lead to better resource productivity and more innovation and ultimately to increased competitiveness. Reducing pollution and limiting waste can improve efficiency and reduce costs. Complying with environmental regulations and responding to external pressures can increase innovation resulting in better products, improved product yields and improve global competitiveness.

11.4   Strategic direction

Strategic flexibility

Hitt et al (1998) in discussing the way in which the competitive business scene is changing (with increasing globalisation and technological change) suggests that organisations need to build and maintain competitive advantage. This requires the strategic flexibility to respond quickly to changes and the strategic leadership to achieve and sustain the response. Table 11.2 summarises the principles needed to build this strategic flexibility and create competitive advantage.

Environmental objectives

Incorporating environmental objectives into an organisation should start at the strategic level when the strategic business objectives of an organisation

Table 11.2 Strategic flexibility and competitive advantage

Strategic Principle	Components
Build dynamic core competences	• Unique set of resources built into skills and capabilities
Focus and develop human capital	• Developing employee skills • Contingent workers & outsourcing
Effectively use new technologies	• Manufacturing technologies • Information technologies
Engage in valuable strategies	• Use co-operative strategies • Exploit global markets
Develop new organisation structure & culture	• Horizontal organisation • Learning & innovative organisation culture

Source: Based upon Hitt et al (1998), p. 26.

are being developed or revised. Dunphy et al (2003, p. 171) discusses how environmental aspects can be integrated into the framework of Table 11.2 by adding human and ecological sustainability to define a concept of strategic sustainability. Below we discuss some of the global and local forces that need to be considered when establishing environmental strategic objectives.

Global environmental and social issues

Outside the immediate neighbourhood of an organisation, there are significant global issues which are becoming more evident and more widely discussed and are filtering their way down to business corporate strategy. These include:

- the unsustainable use of the world's resources (Chapter 2) and the development of the concepts of sustainable development covering social, environmental and economic;
- global initiatives to address this problem through global mechanisms, global organisations and peak business groups;
- the continuing disparity between the developed and underdeveloped countries i.e. intragenerational equity;
- the needs of future generations i.e. intergenerational considerations;
- growing globalisation of industry and with it global pollution and resource use and the need to consider local, regional and international environmental issues; and
- the increasing pressure for business organisations to contribute to the solution of the social and environmental problems of the world.

External forces for change

The early chapters of this book provide a background to some of the external forces that are driving business organisations to reconsider their corporate strategy so as to develop a complementary environmental strategy. There is a need to respond to:

- government and legislative demands concerning pollution, waste, reporting (voluntary/statutory) and conditions of approval for developments;
- community and green group demands, complaints, protests at the environmental and social impacts of their operations concerning use of resources, manufacturing, waste, product disposal and stewardship;
- consumers' needs and wants concerning green products, 'organically' grown food, degradable packaging;
- the growing market for environmental goods including pollution control equipment, environmental consulting services; and
- the push for more accountability to government, neighbours, shareholders and owners.

Advantages of responding environmentally

There are potential advantages to business by responding positively and proactively to the environmental challenge. They include:

- achieving *legislative compliance* and maintaining permission to operate;
- *reducing costs* by good environmental management especially in the more efficient use of materials and energy;
- achieving *product differentiation* by producing more environmental friendly products and services;
- achieving *initial gains* by being first into the green market i.e. first mover advantages and collection of 'the low hanging fruit'; and
- achieving *improved performance* with a more sustainable operation and committed workforce (Welford, 1998; Porter and van der Linde, 1995).

And as Porter says (Section 11.3) this response can lead an organisation to more innovation and ultimately to increased competitiveness.

Most of these aspects can be felt at the strategic level of an organisation and it makes sense for an organisation's business strategy to be integrated with its environmental strategy and for this to link to an Environmental Management System.

11.5   Strategic choice and environmental objectives

Environmental targets for individual process or products are being set by many organisations but at the strategic level strategic medium and long-term targets should be considered. If business organisations are to become part of achieving global sustainability they need to consider long-term environmental targets for their businesses e.g. a ten-fold increase in efficiency in the way they use energy, natural resources and other materials (Hawken et al, 2000, p. 11) and then, or in parallel, move to more rigorous targets like 'zero discharges, no net energy use or use of only renewable energy'.

Choosing strategic environmental objectives is an ongoing process. We do not suggest any in this chapter as each business is different but the following chapters of this book provide the techniques and tools to help organisations to choose, evaluate and implement different strategic environmental objectives.

11.6 Evaluation – measuring environmental strategy/ performance

There are many ways of measuring achievement of strategic environmental objectives and these include the use of: phase or *stage models* which allow an increased strategic performance in time to be measured mainly qualitatively though they can be supported by quantitative measures; *typological models* normally a 2 by 2 matrix that allow an organisation's relative position at one point in time to be assessed against two performance axis but without a time scale; *strategic performance measurement* of which the Balanced Scorecard is a good example; and *environmental performance evaluation* based on combining a sometimes complex array of performance indicators (Kolk and Mauser, 2001).

Stage models, typological models and strategic performance measurement originate in corporate strategic management approaches but are being adapted to include environmental aspects. Examples of *stage models* and *strategic performance measurement* are discussed below. The reader is directed to Chapters 15 and 16 for a more detailed discussion on *environmental performance evaluation* using environmental indicators. Kolk and Mauser (2001) provides an overview of *typological models* and their references sources should be consulted for more details.

Phase models (see also Table 18.1)

There is a frightening range and number of stage or phase models (and typological models) described in the literature that have been designed to allow an organisation to establish strategic objectives and targets and track progress towards them. Kolk describes 49 models related to environmental measurement but cautions in their use. They are generally 'broad and conceptual – which is their main contribution' but many focus on environmental management rather than on environmental performance (Kolk and Mauser, 2001). Readers are encouraged to explore these and assess their relevance for their particular organisation.

Table 11.3   Organisational change for corporate sustainability

No	Phase in corporate sustainability	Measurement Characteristics used for each phase	To achieve ecological sustainability, the business will:
1	Rejection	• Vision and goals • Change agents	
2	Non-responsiveness	• Corporate policy/strategy	
3	Compliance	• Structure/systems • Stakeholder relations • Human capabilities	• Actively promote ecological sustainability.
4	Efficiency	• Ecological capabilities • Tools/techniques	• Espouse & enact environmental best practice.
5	Strategic proactivity	• Production/service systems	• Assist society to sustainability through all its products & services.
6	The sustaining corporation		• Use its influence on government, markets and community values to promote sustainability. • Value nature for its own sake.

Source: Dunphy et al (2003, p. 22, p. 296).

Dunphy et al (2003, p. 22) has developed a phase model that can be used to measure strategic organisational change as an organisation progresses towards a sustaining corporation. Measurement of achievement is through six semi-quantitative mileposts that enable an organisation to assess how it is progressing towards both human sustainability and ecological sustainability. Table 11.3 shows the phases, the measurement characteristics used for each phase and for Phase 6 the aspects of ecological sustainability.

Strategic performance measurement

The Balanced Scorecard is a management and measurement tool used to guide the implementation of corporate strategies and continual improvement. It links financial and non-financial corporate activities to the organisation's long-term strategy through four perspectives: *financial* – how do we look to shareholders; *customer* – how do customers see us; *internal business* – what must we excel at; and *innovation and learning* – can we continue to improve and create value? For each perspective, goals are set and indicators developed and measured as shown in Figure 11.1. The measures and goals are linked to the long-term strategies through cause and effect relationships (Kaplan and Norton, 1992).

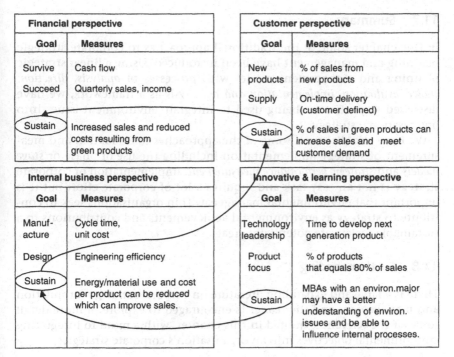

Figure 11.1 Sustainable Balanced Scorecard, example with environmental goals
Source: Based on Kaplan and Norton (1992); Figge et al (2002).

Because the Balanced Scorecard makes it possible to take into account non-monetary strategic success factors that impact the financial success of a business, it is a good starting-point to also incorporate environmental and social aspects into the management system of an organisation. Including sustainability within the Balanced Scorecard can help overcome the shortcomings of conventional segregated approaches to environmental and social management systems by integrating the three parts of sustainability (financial, social and environmental) into a single strategic management tool. It is necessary though to link environmental performance to specific corporate strategies (Figge et al, 2002). One key aspect of the Balanced Scorecard is the need to establish cause-effect relationships that link changes in one area of an organisation to its overall performance (Kemp, 2004).

Figure 11.1 shows an environmental goal for each perspective under the heading *sustain* with its cause-effect links through the matrix to the financial perspective. The innovative trigger is for an organisation to employ people with Masters of Business Administration with environmental majors.

11.7 Summary

In this chapter some of the traditional approaches to corporate strategic planning and management have been introduced. Using a linear strategic planning and management model with processes of *analysis, direction, choice, evaluation, implementation and reanalysis* as a framework, we have discussed ways that are being used to integrate environment issues into organisational strategy.

We have also discussed some of the approaches to evaluation and measurement of strategy implementation including the use of phase or *stage models* and *strategic performance measurement*. Implementation of corporate strategy is not an easy task and requires a lot of corporate effort but it is important that environmental managers within organisations seek to contribute to strategy as environmental achievements and contributions to a sustainable society are potentially great.

11.8 Further reading

There is a large and expanding literature on corporate strategy formulation and management and the reader is encouraged to explore in more detail some of the references quoted in this chapter, with a mind to integrating environmental objectives into an organisation's corporate strategies.

11.9 Questions

- Using the framework in Table 11.1 identify how each of these strategic processes are undertaken in your organisation and whether significant environmental considerations are a part of these processes.
- Using the five principles of Table 11.2 identify approaches your organisation could adopt to set environmental strategic direction.
- Using the phase model in Table 11.3 (and Dunphy et al, 2003) assess in which phase your organisation lies.

12
Environmental Management Systems
Robert Staib

12.1 Introduction

This chapter discusses Environmental Management Systems (EMS) and their function within organisations. An outline is given of the structure of a system, some of the types of systems available and the international standard ISO 14001 and its application. An EMS enables an organisation to manage environmental issues in a structured way. Within its framework many of the issues discussed in this book can be addressed logically. This section is mainly written for an organisation establishing an EMS but because environmental management is ongoing, the processes apply equally to periodic review and continual improvement of the EMS. For more detail on EMSs and how to implement and operate, the reader is referred to the references cited in this section and in particular to the ISO 14001 series of standards. (ISO, 1996a, 1996b; Welford, 1998; Hutchinson and Hutchinson, 1997). Note that ISO 14001 and 14004 have been modified (ISO 2004; ISO 2004b) although many organisations could be still operating under the older versions for a period.

12.2 Corporate governance and management systems

Organisations both public and private need management systems to function and to govern themselves. A typical management system is the accounting system used to track and manage money movements and finance requirements. Other systems include management information systems that provide information on the operation of an organisation that enables the staff and managers to assess the performance of the organisation and make decisions. Systems use information generated by the people within the organisation and by external sources. This is used by the people within the organisation (e.g. costs of production) but can also be provided to external parties for their use (e.g. income for Government taxation calculations).

Over recent times there has been a proliferation of different sorts of management systems within organisations either as stand-alone systems (e.g. a project management system to manage cost and time aspects of projects) or systems that link and integrate existing ones (e.g. quality assurance systems). There has been a codification of the requirements of many of these systems through international codes e.g. the International Organization for Standardization (ISO Standards) or standards particular to an individual country. Standard international management system codes include environmental – ISO 14001, quality – ISO 9000 and national codes for occupational safety and health – Australian Standard 4801 or British Standard 18001. Many of these standards were established to address particular issues within organisations. Because of the proliferation of these standards and their similar management process requirements there is a growing movement both within the standard developers and within organisations to integrate these diverse systems into a single or at least a unified management system (Buck, 2001; Mattieu, 2003).

A management system consists of documented policies, processes and procedures promulgated throughout an organisation. People are given both formal and on the job training in its use and various equipments are acquired to support the system e.g. for monitoring, measuring, collection and storage of data often with the support of computer hardware and software. Management systems are tools to help an organisation achieve its corporate objectives e.g. make a profit, deliver a project on time and budget or maintain a healthy work environment. Effective management systems are not static stand-alone systems but are operated by people within organisations. The people aspects of environmental management are discussed in Chapters 17, 18 and 19.

In this chapter we concentrate on outlining the requirements of an Environmental Management System (EMS) but also discuss some of the issues of integration.

12.3 Why have an environmental management system

Environmental management systems (EMS) are a way of enabling an organisation to address and manage the plethora of environmental issues that it is faced with – originating from society or from within the organisation. A well run EMS can provide an organisation with a number of benefits:

- a structured and systematic approach to compliance with environmental legislation;
- a legal protection against prosecution;
- a tool to identify and meet future environmental and legislative change;
- a clearer definition of the environmental values of the organisation;
- an approach to management and monitoring of environmental performance;

- better documentation and data management;
- provision of information necessary for internal and external environmental reporting;
- support for an improved market image; and
- a basis for taking the organisation to a performance beyond compliance and maybe towards sustainability (Harland, 2003; Welford, 1998; ISO, 1996a).

12.4 Environmental management system objectives and principles

The fundamental objective of an EMS is to enable an organisation to comply with the current and impending legislation. This is the compliance objective e.g. some legislation now makes pollution a criminal offence both for an organisation and for its members, so compliance is required from both the organisation and its individual members, particularly its directors and key managers.

Other immediate objectives may include improving the environmental performance of the organisation in reducing pollution, minimising the use of resources and protecting the natural environment. After an EMS is established and the organisation starts to integrate environmental issues into its general business, it can establish longer term objectives such as the elimination of all pollution, eliminating all waste, using only renewable resources and enhancing the natural environment (Dunphy et al, 2003) .

Management systems can tend to become complicated and can require much effort in their implementation and operation and become ends in themselves. When all the detail of an EMS is in place in an organisation it is worth reminding oneself that the EMS is only a management system to assist the organisation in managing environmental issues and is a vehicle to help achieve environmental goals. For the system to be beneficial for the environment the organisation should set real and achievable environmental goals that can be steadily enhanced over time. A goal of 'reducing carbon dioxide emissions by 5% per annum' is a more positive goal and more likely to bring benefits to the environment than a goal of 'minimise greenhouse gas emissions'.

Types of environmental management systems ISO, EMAS, CARE

The need to address environmental issues commenced in earnest in the years after the Second World War with the growth in industrialisation and its consequent pollution. Many authors over the last 30 years have described how increasing world population and increasing use of energy with waste producing technology (especially in the period of rapid economic growth since the Second World War) have contributed to the production of a greater quantity of pollutants discharged into the environment (Chapter 2). These pollution problems have been significant for industrialised countries (Section 3.2). The factors that contribute to the rise in pollution (population increase, industrialisation and energy use) have been increasing steeply since World War II (Chapter 2).

After initial adhoc approaches various structured and standard processes and systems started to appear. British Standard BS 7750 was one of the first to appear. It became the basis for the ISO 14000 standard. Other systems include the European Union standard Eco-Management and Audit Scheme (EMAS) for companies performing industrial activities with only discrete sites being able to be registered (Welford, 1998). Various industry schemes and systems have been established e.g.. Responsible CARE is an environmental management process used by the chemical industry (Section 10.4).

Environmental management system under ISO 14000

The standard that is being adopted world wide ISO 14000 is a series of codes covering environmental management systems, auditing, environmental labels and declarations, environmental performance evaluation, life cycle assessment and environmental risk management. By mid 2003, 37,000 organisations world wide had been formally certified to the standard ISO 14001 (Bryden, 2003). An organisation can use this code as a basis for its own system or it can adopt all parts of the code and obtain and maintain certification by an external accreditation body. In the following sections we describe and discuss the main components of an EMS developed under ISO 14000.

Structure of an EMS

ISO 14000 is based on a simple management process model with a feedback loop to ensure that established goals or objectives are achieved by measuring performance and regularly reviewing goals and processes. Figure 12.1 shows

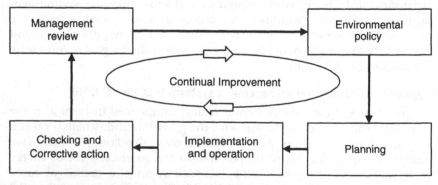

Figure 12.1 Environmental Management System model
Source: Based upon the figure from ISO (1996a).

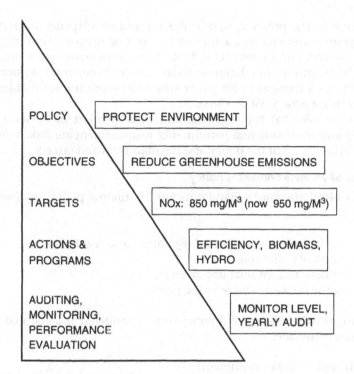

POLICY

OBJECTIVES

TARGETS

ACTIONS & PROGRAMS

AUDITING, MONITORING, PERFORMANCE EVALUATION

PROTECT ENVIRONMENT

REDUCE GREENHOUSE EMISSIONS

NOx: 850 mg/M^3 (now 950 mg/M^3)

EFFICIENCY, BIOMASS, HYDRO

MONITOR LEVEL, YEARLY AUDIT

Figure 12.2   Hierarchy of EMS elements, e.g. a coal fired power station company
Source: Based upon ISO (1996a); Delta Electricity (2000).

this model with six processes: environmental policy, planning, implementation and operation, checking and corrective action, management review and continual improvement.

Hierarchy of EMS elements

Another way of visualising the EMS is through a hierarchy of elements starting with the organisation's environmental policy and then proceeding to the detail of objectives, targets, actions and programs, auditing, monitoring and performance evaluation. This enables an organisation to relate the detail of everyday operations to the organisational environmental goals established broadly in the policy and increasingly more specifically in objectives and targets. Figure 12.2 shows the hierarchy and how it might apply to a typical coal fired power station.

12.5   Defining an environmental policy

The environmental policy outlines an organisation's commitment to environmental management. It should also be a commitment to environmental achievement. The policy should be flexible and be adaptable to changing circumstances and changed achievements and be related to the corporate vision, mission and values (Table 11.1). It is a statement by the organisation of its intentions in relation to its overall environmental performance and provides a framework for the setting of its environmental objectives and targets for action (ISO, 1996a).

It is desirable that the organisation's measurement processes e.g. its auditing and environmental performance measurement are linked directly to the achievement of the policy and the objectives and targets.

Content of an environmental policy

The ISO 14001 standard's mandatory environmental policy requirements are:

- compliance with environmental legislation and regulations;
- prevention of pollution;
- commitment to continual improvement; and
- making the policy available to the public.

Other items of environmental policy to be considered and adopted by an organisation include:

- its attitude to the environment;
- establishment of quantified environmental goals;
- commitment to meeting and exceeding environmental standards;
- conservation of natural resources;
- minimisation of the effects of its products and services;
- provision of a healthy workplace;
- liaison with local communities and society;
- training of staff and suppliers;
- addressing all environmental issues e.g. energy, waste, land, water, air;
- commitment to review of environmental performance; and
- commitment to moving towards an ecologically sustainable company (Dunphy et al, 2003) and achieving no net environmental impact.

Some further important aspects of establishing and maintaining an environmental policy are:

- obtaining endorsement from top management;
- making it appropriate to nature and scale of impacts of an organisation;
- making a commitment to comply and achieve the policy;
- using it as a framework to set real quantitative environmental objectives & targets; and
- making sure it is documented, implemented, maintained and communicated.

12.6 Environmental planning

Planning (both preliminary and later detailed planning) is important in establishing and maintaining a comprehensive and robust system. The first step in establishing environmental goals is to identify the major environmental issues. This can be done by listing the major business activities and the main products and services of the organisation. For each of these items the environmental aspects and environmental impacts are listed (See example in Table 12.1). The impacts can then be quantified to assist in the evaluation of their significance with the most significant impacts being addressed initially (ISO, 1996b). A number of tools that can be used to help identify, assess and prioritise the environmental aspects are presented in Part IV e.g. project environmental impact assessment, cleaner production, life cycle assessment and risk assessment.

A detailed study needs to be made of all legislative requirements. This will cover the legislated Acts, regulations and Statutory authority policies (that support the Acts) and various permits, licences and approvals (that an organisation either has in place or is required to obtain). At this point it is also useful to identify impending legislation and other potential legal

Table 12.1 Environmental aspects and impacts – example

Service	Aspect	Issue/impact	Program & function
Facilitate urbanisation	Water quantity	Flooding from urbanisation	Dry basins on creek mitigate flooding, but impact creek continuity
Supply drinking water	"	Potable water use	Recycled water reduces potable water demand
Facilitate urbanisation	Water quality	Urban runoff pollution	Wet basins trap nutrients and prevent pollution, but impact creek continuity
Remove waste water	"	Sewage pollution	Tertiary treatment protects creeks, recycled water reduces amount of effluent.
Reticulate water mains	Land use	Clearance of flora	Landscape repairs of disturbed areas
Reticulate water mains	"	Damage to archaeology	Archaeological sites are salvaged scientifically and culturally

Note: Example from a water supply corporation (Staib, 2003).

requirements. The mandatory level of legal compliance should be separately identified including the requirements for disclosure and penalties the organisation and its staff would incur for non-compliance.

Good planning leads to the setting of realistic environmental objectives and targets. To achieve these targets it is necessary to develop implementation programs. The programs could be one off projects, could be part of an ongoing research and development or production. (Refer to Chapter 25 Environmental Impact Assessment, Chapter 22 Cleaner Production.) Other aspects that should be examined at this planning phase include existing environmental management programs, management practices and investigations of previous environmental incidents, reports and audits.

Environmental aspects

The environmental aspects and impacts should include (ISO, 1996b; Staib, 1997–2000):

- emissions to air including noise;
- releases to water and land;
- contamination of land;
- waste reduction and elimination;
- impact on communities and stakeholders;
- impact on heritage – natural, cultural and built;
- use of raw materials and natural resources;
- use of energy – process and embodied;
- impact on natural systems and biodiversity;
- impacts of contractors and suppliers;
- impacts of users of the organisation's products and services; and
- cumulative impacts and long-term sustainability issues.

Setting realistic quantitative targets

It may not be possible at the start of a program to immediately set quantified (and measurable) objectives but the aim should be to do so in the immediate term so that they can be monitored and reported against. The objectives should include indicators and target values for the indicators for the main impacts of the organisation. They should be focused on reducing the environmental impact of the organisation.

12.7 Implementation and operation

The implementation and operation of an EMS is similar in many respects to other corporate systems and requires adherence to the principles and processes of general organisational management. Many organisations have quality systems (Bryden, 2003) and those familiar with them will understand some of the terms and processes of an EMS. For environmental management to be accepted by people at all organisational levels and to be successfully implemented by an organisation it needs to be integrated into the general running of the business. For many people it will mean an added responsibility though they may be supported by specialists providing advice and information and performing specific tasks.

The ISO 14001 standard (ISO, 1996b) requires the following aspects to be addressed: structure and responsibility, training (awareness and competence), communication , EMS documentation, document control, operational control, emergency preparedness and response and supply chain management. These aspects should be considered in concert with existing systems within the organisation and duplication and separation avoided. If possible environmental aspects should be extensions of existing systems and roles integrated.

The key aspects concerning the organisational issues that need to be defined and implemented for an EMS are:

- the existing organisational structure and personal responsibilities need to be reviewed and augmented to cover environmental responsibilities;
- the roles, responsibilities and authorities for each organisation position need to be clearly defined and documented;
- the organisation needs to provide adequate resources – people, technical, financial; and
- one person should have ultimate responsibility for the EMS (Welford, 1998).

From both environmental and human points of view it is important that each person in the organisation is given an environmental responsibility however small and their commitment sought. The Board and Chief Executive should have overall responsibility for without a commitment and leadership at this level environmental performance will be poor and hope of achieving considerable reduction in environmental impacts remote (Dunphy et al, 2003; ISO, 1996a).

Training, awareness, competence

The organisation needs to identify and provide appropriate training for its people. All people should receive some environmental awareness training covering key aspects e.g. general environmental issues that impact society, the issues that concern their organisation, the use of the organisation's EMS, relevant legislation and external approvals that cover their work, role and responsibilities. In addition some people will require specific skills training and some jobs will require recognised external qualifications and considerable experience.

Not all skills will be available in-house and organisational resources may need to be supplemented by consultants or recourse to the information and services provided by industry peak groups. This might especially apply to small and medium sized enterprises (SME).

Communication

Communications within the organisation for receiving, acting on and distributing environmental information should be established as well as communications with external stakeholders and internal stakeholders. Some authors (Dunphy et al, 2003; Welford, 1998) see the organisation as having an obligation to serve society rather than focussing merely on the interests of directors, owners and shareholders. Other stakeholders are increasingly demanding that organisation become more socially and environmentally responsible (Chapter 9). Therefore communications with key external stakeholders e.g. suppliers, customers and the general public are important.

Environmental management system documentation

This could be in the form of an EMS manual with description of each element and maybe a supplementary volume with environmental management procedures, standard forms and reporting formats. If a company has a certified Quality Assurance system the framework for an EMS manual could be already established. Integration of the EMS with other corporate systems e.g. occupational health and safety, quality management could be attempted after the EMS is operational. Having a separate EMS initially will help people focus on the unique aspects and requirements of the environment.

Operational control

Operation control will cover the procedures above and their implementation and management. The procedures should be structured to ensure that the objectives and targets are achieved and the environmental policy is adhered to.

Emergency preparedness and response

Organisation should have in place procedures to cover environmental emergencies and incidents to mitigate potentially significant impacts. These procedures need to identify potential accident and emergency situations, document a planned response to these situations if they arise and prevent and mitigate any environmental impacts associated with them. They should be integrated with the organisations traditional emergency procedures so that when an emergency or incident occurs the environmental aspects are not forgotten. A suggested hierarchy of responses for an integrated system would be: personal safety, environmental impact and then asset protection though in many situations these may be linked.

12.8 Checking and corrective action

The next two sections are the management feedback loop of Figure 12.1 and are easy to ignore in the heat of the battle. They are important to ensure that progress towards agreed targets is monitored and that the appropriateness of the targets is continually assessed. The checking and corrective actions provide information to the organisation to help it to achieve targets and make ongoing improvements to take it down the path to sustainability. This process includes: monitoring and measurement of environmental indicators, identification of non-conformances, instigation of corrective and preventive actions, keeping of records, undertaking environmental management system audits and process/product audits. To resolve non-conformances and to learn from mistakes it is important to ask the question, 'Why has this non-conformance occurred and what can we do to prevent it happening again?'

The EMS should contain procedures for checking and corrective actions and a schedule or programme to ensure the actions take place regularly.

12.9 Management review and continual improvement

Nothing is sacred in business with change being the order of the day. Organisations need to respond to change if they are to stay viable and, if they are to become and remain sustainable in all the senses – functionally, financially, socially and environmentally (Dunphy et al, 2003). Regular management review of the EMS is a part of this need and will assist the organisation to achieve continual environmental improvement and to ensure that the EMS remains appropriate for the organisation and the environment.

These reviews need to be held regularly, attended by key decision makers and formally documented with clear actions and responsibilities. They need to be related to the business and environmental needs of the organisation. They would typically address the results of checking and corrective actions e.g. audits, internal reports, external reports, potential changes in legislation, changes in the organisation's markets, changes in community and political expectations.

Integration into corporate management processes

The benefit of a separate EMS is that it focuses attention on environmental problems but there could be a tendency to see the Environmental Manager and his/her staff as being solely responsible for environmental performance rather than the whole organisation. On the balance the author believes that it is best to have environmental aspects integrated into the normal cut and thrust of the business of the organisation (Buck, 2001).

One obvious area of integration is with similar systems e.g. Quality Assurance systems (under ISO 9000) and OHS (under BS 18001 or AS 4801).

The International Organization for Standardization has codes that address this integration and identify areas of overlap. There is also the potential to integrate the environmental aspects into the traditional business processes e.g. strategic planning (Chapter 11), value chain management, marketing (Chapter 13) and accounting (Chapter 14).

Computerised environmental management systems

Some organisations, particularly large ones with diverse operations are implementing computerised EMSs. This has advantages in making the relevant information consistent and readily available, though it does mean that the information has to be continually updated. This is particularly important with the latest legislation and authority approvals to ensure basic compliance. Computerised EMSs can also encourage participation of all employees by making them aware of the organisation's environmental goals and how they are being achieved.

12.10   Conclusion

Some criticisms of EMSs developed under ISO 14000 are that they are: too bureaucratic and cumbersome; focus on system and obscure focus on real environmental outcomes; head off government action; too hard for SMEs to implement; part of the 'greenwash' surrounding sustainable development and the Standard was drafted and established by a 'club' and has produced an industry for standards consultants. Notwithstanding these it is being adopted by many organisations both private and public in many countries around the world. In June 2003, 37,000 organisations had externally certified systems (Bryden, 2003). Many organisations are now being required to have EMSs to continue to operate in certain markets e.g. for government work, for major companies (the car maker Ford's suppliers).

Recent work indicates that for EMS to be successful human resources factors such as training, empowerment, teamwork and rewards also need to be successfully addressed (Daily and Huang, 2001). These issues are discussed in Part III of this book. Harland (2003) believes that the following system and human factors are critical to the success of implementing an EMS:

- a system tailored to the needs of the organisation;
- the setting of realistic and attainable goals;
- maintaining procedures that are simple, clear and concise;
- involving those using the system;
- achieving commitment from top management; and
- adopting a non-punitive approach with open honest two-way communications.

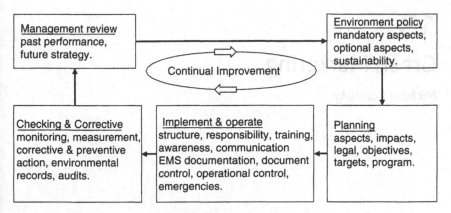

Figure 12.3   Environmental Management System elements

Figure 12.3 summarises the structure and the elements of an EMS that we have discussed in this chapter.

12.11   Further reading

Latest versions of ISO standards 14001 and 14004: ISO (2004); ISO (2004b).

12.12   Questions

- Choose an organisation and prepare a report to the Board of Directors recommending (a) the introduction of an EMS or (b) if one exists, its substantial (or transformational) change to embrace sustainability (See Chapter 18).
- Discuss some of the pros and cons of introducing an externally certified EMS to ISO 14001 into an organisation of less than 20 people.

13
Green Marketing

Michael Polonsky

13.1 Introduction

This chapter examines the inclusion of environmental issues in marketing activities. There are a variety of ways that this can be achieved and the degree to which environmental issues are included in marketing activities can vary. As such this chapter briefly discusses many of the most relevant issues and provides some examples of how business organisations from around the world have undertaken these activities. In all successful cases organisations have made meaningful environmental improvements that are then leveraged in marketing related activities.

13.2 What do we mean by green marketing?

The definition of green marketing is illusive. At the more general level authors such as Polonsky (2001b) has suggested that Green Marketing – is an attempt to minimise the negative environmental impact of production and consumption. Others such as Menon and Menon (1997) have suggested three levels of greening:

- *Strategic Greening* – requires a fundamental change in corporate philosophy e.g. an organisation redesigning its whole operation to be a closed loop system, with minimal waste;
- *Quasi-Strategic Greening* – requires a substantial change in business practice e.g. hotel chains instituting procedures to reduce water consumption associated with washing linen, where guests are asked to indicate whether they want their towels and linen washed daily; and
- *Tactical Greening* – with shifts in functional activities, such as promotion e.g. an organisation undertaking a promotional activity linking their products to environmental issues, although there is no change in corporate activity overall.

From a marketing perspective, when is an organisation green? Is an organisation that makes one 'environmental' product (alongside other traditional goods) targeted at a very narrow market segment really practising green marketing? The answer depends on the definition used. One would have to say that it is better that this organisation is doing something. However, this organisation might simply be operating opportunistically and not really reducing its environmental impact.

While in an ideal world all organisations would see green marketing as a core component of all activities, this is not the case. Each organisation will need to consider the degree to which it greens its activities. Organisations that make green products will need to consider the degree to which green information is actively communicated to consumers. In some cases organisations may have substantial environmental improvements that are not communicated, simply because the organisation is concerned with consumer scepticism (Peattie, 1999). Organisations also need to consider how greening activities are perceived by consumers and other stakeholders, as in some cases consumers perceive green products to represent inferior value (Levine, 2002). It is also important to consider how greening products will impact price? E.g. it is possible to 'recycle' tyres back to the constituent components (Bureau of International Recycling, 2004), although it may not be economically viable.

Green marketing sometimes is seen as a double edge sword. Promoting oneself as green might result in the organisation being scrutinised more heavily (Levine, 1993; Frost, 1991). Unfortunately in some cases where an organisation has made substantial improvements, it might be judged more harshly when problems arise, than other organisations not making the effort.

From the discussion of Consumer behaviour in Chapter 9, 'green' consumers can be considered in a number of different ways ranging from ecologically responsible (those who evaluate the environmental impact of consumption, but still consume) to simplifiers (those who seek to reduce their mass consumption and minimise their impact on the natural environment). Different groups will view environmental issues differently and thus need to be evaluated separately by the organisation (Connolly et al, 2004).

13.3   Green markets

The greening of marketing activities can be applied to a diverse range of markets: consumers, businesses, governments and investors. Each of these markets can be targeted in different ways, as each may have different environmental motivations and goals.

Consumers

Consumers most frequently consider a product's environmental attributes because of some special interest in the environmental issue. That is, they incorporate environmental criteria as an attribute to be considered alongside other attributes. While there are some consumers for whom green issues drive the discussion process, these may tend to be in the minority, e.g. Ottman (1998) suggests that there are different categories of green consumers behaving differently. One needs to understand where the majority of one's consumers fit and how to market to them.

Suppliers

With an increasing interest in environmental activities, many organisations are seeking suppliers and consultants that can assist them in improving the eco-efficiency of their activities e.g. green issues can be considered and integrated when designing new products and processes, with eco-design being an important component of overall design (Bhat, 1993, Fuller, 1999).

Business and government

The inclusion of green marketing for businesses and governments may in some cases be an operational necessity rather than an optional activity. Organisations that have adopted International Standard 14001, often require their suppliers to comply with ISO 14001. While there has been some suggestion that ISO 14000 and other standards could potentially restrict trade (Dohlman, 1990), ISO 14000 accreditation could serve as a value tool in marketing to other businesses and/or governmental bodies seeking specific environmental performance levels. This is not to suggest that organisations that do not have such accreditation do not meet these standards, but rather have chosen not to seek accreditation.

Governmental bodies by mandating supply and demand can create market opportunities. Governmental bodies can regulate their own purchases to ensure that minimum environmental standards are met e.g. they can require all paper products to have some minimum level of recycled content (Jacobson, 1993). Governmental bodies can also require that producers meet specified environmental standards e.g. the state of California at one time had a policy that X% of all cars had to be electric; in turn motivating auto manufactures to respond (Byrne and Polonsky, 2001; Eaton, 1995).

13.4 Greening the marketing mix

There are many ways that organisations can operationalise greening and integrate it into overall strategy (Peattie, 1999). In this section we discuss some of the generalised marketing mix activities (Product, Price, Place, and Promotion). These activities can be undertaken with differing levels of enthusiasm within the organisation: strategic, quasi-strategic or tactical. In addition the greening activities need to be integrated across the marketing mix and other organisational activities (Polonsky, 2001b).

Green products and product design

The incorporation of environmental attributes into products and processes needs to be undertaken at the initial stages of new product development (Bhat, 1993; Sharfman et al, 1997). In some cases solving environmental problems within the production process has generated a whole industry of new products, technologies and services e.g. catalytic converters were developed to reduce automobile pollutants. They are a product modified to reduce the harmful environmental impacts (i.e. end of pipe solution), not to prevent the emissions from being produced (Butler, 2002). More integrated greening occurs when automobile manufacturers develop more efficient automobiles with new technologies that eliminate harmful emissions and make cars that are easily recycled (Lewis, 2002). There are other more innovative ways to consider greening products where the organisation explores new technologies that shift the way products operate rather than how they are produced (Roy, 1999).

There needs to be an integrated design process to ensure that products are less environmentally harmful. Ashley (1993) suggested that decisions made during this stage contribute 70% of a product's environmental harm. One approach for designing in environmental considerations is to undertake a life-cycle assessment (Chapter 24) where the organisation considers production, sourcing, use and disposal issues (Oakley, 1993).

The organisation might be able to develop a product that addresses consumers' needs in a new way e.g. some studies suggest that consumers can have their needs met without them owning products, by sharing or renting need satisfying capacity (Belz, 1999). This could include people and businesses purchasing the right to access transportation rather than buying cars (Ho, 2000). Such changes require that customers recognise that needs can be satisfied equally well without their having to own products.

Green pricing

How should one price green products? A marketing approach would suggest that the price should be consistent with the perceived value, not costs of production (although it needs to cover these costs). In some cases 'green' products have had a premium price resulting from higher production costs, which are passed on to consumers. Higher prices can also represent increased 'value' (real or perceived). On the other hand, environmental products could have lower prices, simply because of efficiencies resulting in reduced production costs (Porter and van der Linde, 1995).

A second issue associated with price is how organisations can get consumers to consider the life-time cost of goods, i.e. purchase price, maintenance and operating costs, all of which should be integrated into the pricing and purchasing decision. Consumers tend to be less holistic when evaluating products. (See the light bulb example in Section 9.4) It is unclear if consumers can be convinced to consider this approach.

While consumers report that they will pay a premium for less environmentally harmful products, practice and research has found that the real premium paid is small –5% to 10% (Kapelianis and Strachan, 1996). When marketing green products it is therefore, essential that the environmental value is important to consumers and this increased value is communicated to consumers.

Greening of place and logistics

Greening of place can be examined as the movement of goods from suppliers to the organisation (traditionally logistics) and the movement of goods from the organisation to consumers. Consumers tend not to be concerned with how products arrive at retailers, but rather that the products are readily available at a reasonable price. Changes to shipping, packaging and physical transportation are activities which often do not impact on consumers' evaluations, but may have substantial environmental impacts.

Reduced packaging materials can lower material costs and distribution costs (i.e. lighter products or less breakage) e.g. shipping pallets are a valuable commodity, which organisations can lease rather than sell (Gooley, 1996), with reduced wood usage, and lower costs for organisations renting rather than purchasing (Auguston, 1996). Such activities have shown that organisations see used shipping materials as valuable (Glenn, 1996).

Increased use of the Internet might reduce the need for multiple distribution centres and the need for retail outlets, with an environmental benefit, although this would have to be balanced against increases in other transportation activities (Schegelmilch et al, 1995; Wu and Dunn, 1995). The Internet could also reduce intermediaries with integrated systems reducing the length of distribution networks (Campbell and Green, 1999).

Reprocessing waste and used materials is another important environmental marketing activity, where organisations recognise that 'used' materials have value that can be 'harvested' (Tibben-Lembke, 1998). In some cases the increased interest in reverse channels is motivated by regulatory pressure, although market forces also impact its implementation (Micklitz, 1992). Although reverse logistics and reprocessing have a range of costs including the reprocessing and 'checking' of goods to meet performance standards (Ayres et al, 1997; Giuntini and Andel, 1995).

Greening promotion

Marketing promotion involves a range of tools: advertising, sales promotions, personal selling and publicity (Kotler, 2003). Greening of promotion is primarily concerned with communicating the environmental benefits of an organisation and its products, with consumers and stakeholders. There must of course be something worthwhile to communicate about. In addition consumers need to be able to understand this information, relate it to the product and also be concerned with this issue. What environmental information should be communicated and how should it be communicated (Shrum et al, 1995) is not an easy question.

Organisations that make false environmental claims will receive negative feedback from consumers and stakeholders who will be quick to 'punish' organisations that mislead them (Newell et al, 1998). In the early days of green marketing this 'puffery' was one of the main reasons for consumer scepticism, which hurt those making meaningless claims as well as those communicating substantial environmental improvements (Carlson et al, 1996; Mohr et al, 1998).

Globally public policy makers have introduced regulations that seek to ensure green promotions are not misleading (Kangun and Polonsky, 1995). Consumers' inability to truly evaluate the environmental attributes of products makes such regulation necessary (Newell et al, 1998) e.g. how can a consumer determine if a product's packaging is 100% post consumer waste? In some cases governments, private businesses and not for profit organisations have sought to set up 'standards' to demonstrate they are acting responsibly, but there is a plethora of such programs (Eco-labels, 2004) and it will be difficult for consumers to understand them all.

A range of different promotional approaches can be applied e.g.
* making environmental claims about the product/production process or organisation claims;
* having goods endorsed or certified by third party organisations;
* undertaking sponsorships of environmental activities; and
* participating in joint promotions with environmental organisations.

All promotional communication needs to be related to an organisation's activities and the environmental issue being described. Programs where there are no real links may be seen to be exploitative and will not be effective in communicating positive messages with consumers.

13.5 Other green strategies

The previous section focused on the greening of traditional marketing tools. There are some broader marketing strategies that can be used, which integrate these tools across organisational activities and may shape overall organisational strategy.

Targeting

While there are substantial numbers of environmentally concerned consumers out there, organisations would generally be ill advised at times to develop products solely targeting green consumers and in fact the number of 'green' consumers seems to change over time (Ottman, 1998). Even the 'deep green' consumers will expect 'green' goods to be competitively priced, as well as perform to a satisfactory level. What this means for organisations targeting green consumers is that a product's level of greenness is often used to differentiate two relatively equal goods, with environmental performance becoming a key point of differentiation.

When targeting a segment of consumers, green or otherwise, the segment needs to have the following characteristics: they are measurable in size; they can be contacted through various marketing tools; they are large enough to be profitable (which will vary based on the goods); and they can and will respond to the marketing offering (Kotler, 2003). Green targeted goods can achieve these objectives and with new tools such as the Internet, it may in fact be possible to more easily access markets. In addition environmentally oriented consumers may in fact actively seek out products and organisations that emphasise environmental attributes or benefits (Peattie, 2001). The use of the Internet may therefore allow organisations to access a wider spectrum of green customers.

Green positioning

The role of environmental issues in terms of the organisation's image is something that needs to be considered early in an organisation's development (Crain, 2000). To what extent is 'green' part of the organisations positioning? Is the organisation one that produces quality goods and services that at the same time minimises environmental harm, or is the organisation environmentally focused but also produces *want satisfying* goods and services? The positioning issue needs to consider how consumers perceive the organisation and thus impacts on purchase behaviour as well as attitudes towards the organisation.

Organisations like the Body Shop position themselves on their environmental attributes (Dennis and Neck, 1998). Whereas organisations such as Toyota and Xerox position themselves on quality products, while still emphasising environmental characteristics to varying degrees (Polonsky, 2001b). Other organisations that have a strong internal environmental emphasis may not leverage their behaviour at all. In this way the question is not just to what extent are environmental values integrated into the organisation, but it to what extent are these values part of the organisation's broader 'image' (Menon and Menon, 1997).

Organisations that choose to have an environmental positioning strategy *must* ensure that they perform in accordance with these expectations and needs to be sure that *all* activities support this image. Deviations from stated organisational values would most likely generate negative publicity, even if it were not within the organisation's control (Peattie, 1999). This might explain why some organisations, which are environmentally focused choose to not use this for positioning purposes. For example, S. C. Johnson has as one of its core-values a concern for the natural environment and has won environmental awards from the United Nations and the US EPA (S. C. Johnson, 2004), yet they do not solely position their products as being 'green'.

Marketing waste

In the discussion of place we introduced the idea of reprocessing packaging and products that are no longer needed by organisations and consumers. The marketing of waste is a slightly different activity as it focuses on identifying uses and markets for materials that traditionally are considered to have no value and possibly even be considered to be a cost of operation. Organisations might want to think of waste as a product of organisational activities and, like all products waste has some value (Reynnells, 1999).

The marketing of waste usually starts with organisations seeking to reduce waste, or reprocess waste for internal consumption (Ottman, 1993). They might then move to looking at alternative uses for 'waste' e.g. wineries might pay to have their post-production waste collected, processed and bought back as fertilizer (Tom, 1999). New industries might even be developed to process and market these waste-based goods.

Green alliances

Green organisational partners can assist in developing less environmentally harmful products by identifying new approaches, products and processes that will enable the organisation to better address various environmental issues, by adopting a more responsible philosophy (Crane, 1998; Hartman et al, 1999; Lober, 1997, Mendleson and Polonsky, 1995, Polonsky and Ottman, 1998). Part of the marketing value of these alliances is that the credibility of green partners is 'transferred' to the organisation, although the alliance partners also assist the organisation in improving its environmental activities.

To be environmentally effective organisations need to be able to open themselves up to new ideas and new ways of thinking, which is sometimes difficult. There might also be some difficulties in collaborating with green groups when they have different objectives (Hartman et al, 1999) and a green group's credibility is lost if it simply toes the corporate line. Developing partnerships can also be a time-consuming activity as environmental groups want to ensure that the partner's actions are consistent with the green group's image. Thus green groups might only become involved in activities where they see there is an opportunity to make a substantial impact (Mendleson and Polonsky, 1995).

Table 13.1   Green marketing activities across levels

Activity	Tactical Greening	Quasi-strategic Greening	Strategic Greening
Green products and design	An organisation changes material supplier to one with more eco-friendly processes.	Life-cycle analysis to minimise ecoharm is a regular part of the process.	Fuji•Xerox develops Green Wrap paper to be more eco-friendly from the ground up.
Green pricing	Cost savings due to existing energy-efficiency features are highlighted for a product	Hunter Water changes its pricing policy from a flat monthly rate to a per-unit-of-water-used basis	A company changes its business approach and rents (instead of selling) washing machines to customers, who now pay each time they use the washing machine as opposed to owning
Greening place and logistics	An organisation changes to a more concentrated detergent, reducing package size and weight, lowering shipping costs.	Packaging minimisation is incorporated as part of an organisation's manufacturing review process	A reverse logistics system is put in place by Fuji•Xerox to reprocess and remanufacture photocopiers.
Green promotion	An oil company runs a PR campaign to highlight 'green' practices of the organisation to counter an oil spill receiving negative press coverage.	A company sets a policy that product eco-benefits should always be mentioned in promotional materials	As part of its company philosophy, the Body Shop co-promotes social/eco campaigns each year with in-shop and promotional materials
Targeting	Ads mentioning green features are run in green or environmentally-focused media.	An organisation develops a green brand	An organisation launches a new strategic business unit aimed at the/a green market
Green positioning	A mining company runs a PR campaign to highlight existing 'green' aspects and practices of the organisation.	BP-Amoco redesigns its corporate logo, changing to a sun-based emblem to reflect its view of a hydrogen/solar-based future for the energy industry	The Body Shop not only states that it dedicates its business (and products) to environmental and social change, it proactively engages in related activities and encourages its customers to as well.

Table 13.1   Green marketing activities across levels – *continued*

Activity	Tactical Greening	Quasi-strategic Greening	Strategic Greening
Marketing waste	An organisation improves the efficiency of its manufacturing process, which lowers its waste output	Telstra (an Australian phone company) has internal processes so that old telephone directories (waste) are collected and turned into cat litter products by other companies	A Queensland sugarcane facility is rebuilt to be cogeneration based, using sugarcane waste to power the operation
Green alliances	A company funds a competition (one-off basis) run by an environmental group to heighten community awareness on storm water-quality issues	Southcorp (a leading Australian wine producer) forms a long-term alliance with the Australian Conservation Foundation to help combat land-salinity issues	A company invites a representative of an environmental group to join its board of directors

Source: Based upon Polonsky and Rosenberger III (2001a).

13.6   Green marketing in the planning process

The planning process is based on the premise that the organisation under-stands where it is today and where it wants to go. To be effective these plans need to be appropriately implemented to ensure that what is planned actually happens. Control systems also need to be implemented to evaluate outcomes and modify activities as needed (Kotler, 2003). Green marketing is part of this larger process. The organisation needs to carefully consider, why it is greening activities, which will direct strategy in terms of what activities are undertaken (Peattie, 1999).

To be used effectively green marketing requires an integrated approach involving extensive coordination across functional areas within the organi-sation (Peattie, 1999; Fuller, 1999). This includes functions which 'produce' the organisation's products, those traditionally marketed and those tradi-tionally considered not to have value (i.e. waste), as well as strategic units within the organisation that shape its direction and thus determine whether greening issues are to be strategic, quasi-strategic, or tactical in nature (Polonsky, 2001b). In addition there needs to be input from other functional areas such as marketing, which are frequently not actively involved in environmental strategic decision making. Environmental issues are not simply the domain of an environmental management unit, but become a focus or at least are considered by all functional areas. In strategic marketing planning it is essential that an organisation considers what it wants to achieve with its green marketing actions, both across activities and levels of greening (See Table 13.1).

13.7 Summary and implications for business organisations

It is essential that organisations ensure that *all* green marketing activities are integrated across functional areas, especially if they are used for positioning or promotional activities (Fuller 1999). Organisations should not over emphasise their actions, as organisational exaggerations may have unanticipated negative consequences (Newell et al, 1998). On the other hand, it may not be necessary for organisations to promote all green marketing activities. While from a strategic perspective this may seem to be ignoring opportunities, careful evaluation of overall organisational activities might identify that such opportunities are illusionary, i.e. not all activities support the same environmental focus.

The complex nature of the environmental issues associated with green marketing requires that organisations progress only after they carefully consider all potential ramifications. Table 13.2 lists some 'Do's and Don'ts' of green marketing (Polonsky, 2001b), to assist organisations' green marketing

Table 13.2   Some do's and don'ts of green marketing

DO
- Use strategic green marketing as an opportunity to consider innovative ways of satisfying customers' needs.
- Identify both the short-term and long-term implications of adopting specific green marketing activities.
- At whatever level, strategic, quasi-strategic or tactical understand what is necessary to develop integrated green marketing activities.
- Expect strategic green marketing to be a long-term, ongoing process.

DON'T
- Don't think that all types of green marketing, especially tactical, will generate extensive competitive advantages, at least not in the short term.
- Don't allow marketing hype to over emphasise the true impact of your green marketing activities.
- Don't think that once you make changes the organisation can relax, as green marketing requires continual improvements.
- Don't try to push green initiatives in your organisation simply on emotional grounds, rather communicate in the language of business, as in the long-term both environmental and business benefits can be obtained.

Source: Polonsky (2001b).

activities. Of course green marketing is only one part of an organisation's activities and needs to fit within its overall planning process and overall environmental activities.

Green marketing activities need to be considered in the overall planning process evolving as the organisation evolves. A green marketing strategy is designed to assist the organisation in achieving a current set of objectives. As these objectives change the green marketing activities need to change. There is also a changing state of environmental knowledge, which will impact on product, production and consumption activities. Organisations need to be sure that they monitor where these changes are likely to occur to allow them to adopt more quickly and take advantage of opportunities, while still improving environmental performance.

Case study: Kyocera printers

To position or not as a green product is a key issue for organisations. The decision is largely dependent on the organisation's overall activities, not just on activities in one area, while at the same time considering consumers views on the importance of product characteristics, including environmental characteristics. Kyocera Ecosys' laser printer (Cottam, 1994) when first introduced in Australia, was heavily promoted as a *green* laser printer with important product environmental features: the name; using less energy; having a standby mode; having re-fillable toner cartilages; and being endorsed by a prominent Australian environmental group. The promotion also identified the savings in average operating costs to the purchaser, although this was not the emphasis of the promotion. In this case the environmental benefits were the primary emphasis and the reduced financial costs were a secondary selling point.

It was however realised that most of the business purchasers were more interested in the financial benefits and thought that the environmental benefits were a secondary feature. The organisation ultimately shifted the promotion emphasis to product quality and lower operating costs than competitors. The fact that it was less environmentally harmful than competitors was mentioned but not as the primary *value* of the product. In this way the organisation sought to broaden its appeal from simply targeting green consumers and organisations to targeting all types of consumers. As such they positioned the product as being *main stream* and moved away from targeting a narrow green segment.

13.8   Questions

- Discuss the reasons that an environmentally oriented organisation might choose to not use this orientation to market its products or position the organisation?
- How green is enough to allow the organisation to market its products as environmental?
- How can the organisation make environmental attributes, either of the product or the organisation itself, meaningful to consumers?

14

Financial Management and Accounting

Lorne Cummings

14.1 Introduction

This chapter provides an overview of corporate social responsibility and an introduction to the different types of environmental accounting. An examination of some of the recognition and measurement issues surrounding the incorporation of the environment in financial statements is provided, along with a specific focus on disclosure though Triple-Bottom Line Reporting. The chapter concludes with a case study in environmental accounting on Earth Sanctuaries Ltd.

One of the more critical aspects of environmental management for organisations has been how to financially account for and manage transactions which involve the natural environment. Globalisation has brought a marked growth in global trade. Continuing reliance on fossil fuels and the continued industrialisation and urbanisation of both developing and developed economies, have placed an ever-increasing strain on the natural environment. Private industry has been responsible for much of the impetus for development activity around the world, given limited governmental resources and a general shift toward market-based ideologies as a basis for national development. Whilst technological developments and statutory laws have sought to reduce the negative impact of these activities, there is an urgent need for a complete rethink of how we account for the environment. Given that decisions are more than ever before being based largely on financial concerns, it seems only natural that mechanisms be developed which link organisational decision making on the environment with overall financial performance and position.

14.2 Corporate social responsibility

One of the more difficult aspects from an accounting perspective has been how to incorporate environmental activity into organisational financial statements. Traditionally, the organisation or entity, only considers events or transactions that effect the financial statements directly. However, an organisation does not operate separately from the natural environment. This has led some accounting academics and practitioners to advocate a holistic approach to accounting, which considers financial statement activities beyond that which records activities with other 'legal' entities. Academic and professional literature on the subject has covered a number of areas including financial management, disclosure, classification, and philosophical perspectives.

Early literature on the theory of business organisations held that their purpose was to maximise shareholder wealth. One famous quote on the role of the organisation has come from leading American economist Milton Friedman in an article in *The New York Times Magazine* titled 'The Social Responsibility of Business is to Increase Profits' (Friedman, 1970). He states that 'the cloak of social responsibility, and the nonsense spoken in its name by influential and prestigious businessmen, does clearly harm the foundations of a free society', and that 'there is one and only one social responsibility of business – to use its resources and engage in activities designed to increase its profits so long as it stays within the rules of the game'. Neo-conservative advocates continue to oppose any form of 'corporate social responsibility' (CSR) based on the notion that a business organisation's role is not to provide information to stakeholders who have no direct financial interest in the organisation, as this would breach the 'fiduciary duty' owed to shareholders. Unfortunately, some organisations do not play by the 'rules of the game', and cause social and environmental costs that are 'incurred' and borne by society in general.

As a consequence, there has been a broadening of what society expects from business. Based on the traditional principal/agent accountability model (Figure 14.1), the role of the organisation can be broadened from the 'shareholder' to the 'stakeholder' concept. This model is representative of larger organisations, where there is a separation between the owner and

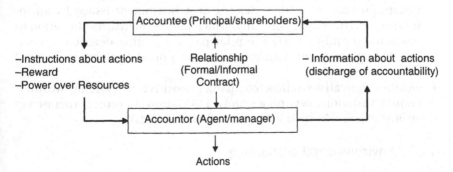

Figure 14.1 A generalised accountability model
Source: Adapted from Gray et al (1996, p. 39).

the manager of the organisation. Under conventional expectations, the primary function of the organisation was to produce profits to shareholders as principals of the organisation. In return, managers as agents would provide audited financial statements, budgets etc, to principals to demonstrate accountability and stewardship of the organisation. This was achieved through formal employment contracts (budgets and other financial information) and legislative requirements (audit reports). Whilst this still remains the case, this model has now widened and in addition to economic performance, shareholders, (in particular those who explicitly integrate moral and ethical concerns into decision making as distinct from just seeking wealth maximisation at all costs) will ask for social and environmental performance criteria. In return, managers provide statements such as 'Triple Bottom Line' (TBL) reports. In the 1970's these were termed Value Added Statements. These voluntary reports are provided to fulfil an informal 'social contract' with society. Although voluntary, over time many of the social and environmental criteria may become mandatory under legislation.

Whilst debate exists as to whether the provision of this information provides benefits in the shorter term, given the costs associated with its collation, classification and presentation, advocates do see longer-term benefits. Apart from responding to contemporary community expectations, disclosure of such information does seek to reduce the risks and costs in an organisation, an example being a reduction in possible legal action associated with social or environmental neglect. Furthermore, a reduction in environmental risk can reduce costs associated with how insurance companies underwrite organisations or how banks lend money. Environmental risk has also spawned new financial products e.g.:

- tradable pollution permits (companies are allocated a pollution quantity permit and if emissions are under the set limit that company can sell the remaining quantity on the open market);
- catastrophe bonds (a high yielding debt instrument issued by an insurance or reinsurance company, which allows the issuers obligation to pay interest and/or repay the principal to be either deferred or completely forgiven when it suffers a loss from a pre-defined environmental catastrophe); and
- weather derivatives (allow companies sensitive to extreme weather/climate variations beyond a standard deviation, to protect themselves against changes in costs and sales of their products).

14.3 Environmental accounting

Environmental Accounting has existed for about 30 years, and can take many forms. To date much of the academic literature has dealt with the social and environmental disclosure in Corporate Annual Reports, including levels of disclosure and motivations for disclosure. Greater emphasis needs to be placed on developing new systems of accounting to record and display such information. The United States Environmental Protection Agency (1995) has classified Environmental Accounting into three levels: National Income, Financial and Management.

National income environmental accounting is where physical or monetary units of measure are used to measure the consumption or utilisation of a nation's resource of minerals, water, arable land, and pollution output – also termed 'green GDP'. In 1993, the United Nations published the 'Handbook of National Accounting: Integrated Environmental and Economic Accounting (SEEA)', which suggested how satellite accounts (e.g. environmental accounts) and alternative classifications, could be included in conventional national statistics. The updated version (United Nations, 2003) includes data on:

- flow accounts for pollution, energy and materials;
- environmental protection and resource management expenditure accounts;
- natural resource asset accounts; and
- valuation of non-market flow and environmentally adjusted aggregates.

Work on formalised systems incorporating the environment has already been used by countries such as Norway for over 30 years, to account for both stocks of forests and fish in their national accounts.

Financial environmental accounting is the process of identifying, collating, and analysing financial and physical information on environmental related activities principally for financial statement purposes and involves:

- *identifying* organisational activities involving the natural environment,
- *measuring* the financial or physical impacts of these activities;
- *recording* the financial or physical costs of these activities on the financial accounts; and
- *communicating* this information in either general-purpose financial reports or special purpose reports.

Management environmental accounting is the process of identifying, collecting, and analysing financial and physical information on environmental related activities principally for internal purposes. Such information is included for more accurate budgeted valuation of products and services in a manufacturing environment, more specifically involving the appropriate allocation of overhead expenses to product costs (Figure 14.2). Previously, 'Hidden Overhead', if not separately identified, would either be included in both Products A and B, or not captured at all, causing not only

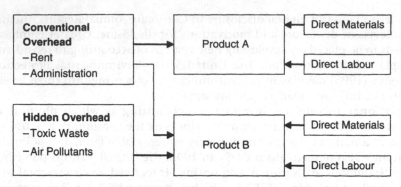

Figure 14.2   Overhead allocation

product mis-pricing, but more importantly a failure to address wastage problems in the manufacturing process.

Identification of hidden overheads such as air pollutants and toxic waste requires there to be management information systems to identify and capture such costs. Organisations have in recent years, been under increasing pressure to adopt international environmental management standards. The primary International Management standard has been the voluntary ISO 14000 series introduced in 1993. The intention is to reduce pollution, by providing an environmental management system framework for organisations to assess their policies and strategies. Peglau (2004) estimates that 835 Australian companies have so far signed up to ISO 14000. Japan at 13,819 has by far the largest amount of registrations of all nations to date, followed by China at 5,064 and Spain at 4,860.

14.4   Recognition and measurement of the environment in financial statements

A considerable challenge for the accounting profession has been to determine to what degree the environment itself can be included on the financial statements. Adopting a purely entity based approach to accounting would limit the degree to which the natural environment can be directly incorporated into financial statements. However there has been an increasing school of thought which adopts a 'holistic' (the philosophical perspective that the whole is more than a mere sum of the parts) approach to accounting. Holistic accounting moves beyond the traditional financial statements to include the human intellectual and natural environmental elements involving aesthetic values and measures.

Assets

Is the environment an asset? Under the International Accounting Standards Committee's (IASC) 'Framework for the Preparation and

Presentation of Financial Statements', *assets* for financial statement purposes are defined as: '...a resource controlled by the enterprise as a result of past events and from which future economic benefits are expected to flow to the enterprise' (para 49).

Whilst items such as land, motor vehicles, plant and machinery satisfy this definition, the environment does not. The environment by its very nature cannot be 'controlled' by an entity. The environment is often seen as a free good that is for the benefit to all. There has often been debate as to the issue of heritage assets. Many of these assets such as national parks are controversial when a valuation is sought on them, as the reliability and relevance of the valuations is called into question.

In 2003, the International Accounting Standard Board (IASB, 2004) determined that Greenhouse Gas Emission Allowances, which are allocated by government or purchased in the market, were to be treated as an 'Intangible' Asset for Financial Statement purposes.

Liabilities

Is the environment a liability? Under the IASC's 'Framework for the Preparation and Presentation of Financial Statements', *liabilities* for financial statement purposes are defined as: '...a present obligation of the enterprise arising from past events, the settlement of which is expected to result in an outflow from the enterprise of resources embodying economic benefits' (para 49).

The environment itself is not an 'entity' and therefore cannot stand before the courts and cannot be a liability unto itself. However liabilities can be imposed in a court of law or via statutory fines, for environmental damage, neglect, or breaches of regulations. Cases often involve proof that the organisation committed an environmental crime, which can be difficult. If an organisation has committed itself to rehabilitation and clean-up costs, or where legal action has been taken against it for environmental damage, then the organisation would record a 'contingent liability' in the notes supporting the financial statements. It would not be recorded as a liability in the financial statements as the cost cannot be 'measured reliably', although it may be possible, in which case the note disclosure would record 'relevant' information for investment decision making.

In 2003, the IASB determined that as greenhouse emissions are made by organisations, a provision/liability is recognised for the obligation to deliver allowances to cover those emissions (or to pay a penalty), which is opposite to the allowance which is to be treated as assets. Table 14.1 indicates the range of possible accounts that may be affected by the environment.

Whilst an identification of the possible range of accounts affecting the environment is an important step, identification of costs remains an important dilemma. Whilst certain costs can be easily classified into one of the

Table 14.1   The impact of the environment on conventional financial statements

Profit and Loss Account	
Revenue	Expense
– Market growth	– Fines
– Market decline	– Health & safety claims
– Product taxes	– Plant depreciation
– Clean-up	– Compliance
– Effluent/emission control or reduction	– Waste minimisation
– Waste treatment/disposal	– Licences/authorisations
– Insurance	– Research & development

Balance Sheet		
Asset	Liabilities	Equity
– Land revaluations	– Capital commitments	– Remediation
– Plant write-offs	– Greenhouse gas	(pollution damage)
– New plant	emission provisions	– Contingent liabilities
– Inventory (NRV)	– Breach of consents	– Natural asset trust
– Greenhouse gas	(fines)	reserve (possible)
emission allowances		

Source: Adapted from KPMG – The National Environmental Unit in Gray et al (1993, p. 23).

above financial statement classifications, often costs associated with business activity are not easily captured in the business statements. An example is pollution. Whilst it is commonly understood that organisations cause pollution, they do not always bear the direct costs of the pollution. The costs of pollution are absorbed by society through lost production, health complaints and a general decline in the quality of life.

As mentioned earlier, the inclusion/non-inclusion of environmental costs onto financial statements can also have an effect on the investment appraisal of an organisation. However it often remains an issue of judgment as to what extent certain social and environmental factors are more important than others. Some important criteria that could be incorporated onto financial statements include:

- the extent of adoption of quality control mechanisms of production;
- past compliance with statutory environmental regulations;
- degree of health and safety requirements;
- prior legal action taken against the organisation;
- third part audits of production facilities; and
- disclosure of physical data on environmental activities.

One of the more fundamental problems associated with accounting for the environment has been how to measure environmental activities. Economic approaches utilise different monetary methods, whilst a physical approach is based on the use of imperial or metric measurements. Economic approaches (Chapter 8) include: *market-price*, e.g. loss on value of contaminated land or crops due to toxic emissions; *hedonic-pricing* attempts to identify how much of a property differential is due to particular environmental differences between properties, e.g. affect of aircraft noise on housing prices; travel-cost uses economic value of 'time' as the indicator of willingness to pay for improvements in environmental quality; and *contingent-valuation* is a survey technique used to estimate an individuals' maximum willingness to pay for a benefit or improvement, or willingness to accept a loss, for a hypothetical issue in question.

The physical approach would record accounts for natural and environmental resources based on physical units. Such measurements would include land used, levels of toxic emissions and biological resources used.

Such approaches are often used when seeking to determine the extent to which a nation's resources of rainforest, mineral etc have been used or replenished over a given time period. The advantages of a physical approach to environment measurement is that it is often easier for an organisation to implement given the ease by which to physically account for depletion/replenishment etc. It also avoids organisations having to place dubious financial measurements which are often unreliable.

14.5 Triple Bottom Line (TBL) reporting

Triple-Bottom Line reporting (Table 14.2) focuses on reporting for the financial, social and environmental aspects of an organisation: *financial* reporting involves traditional value added measures; *social* reporting involves value added that impact upon human and social capital: and *environmental* reporting involves value added measures that impact upon both renewable and non-renewable resources.

TBL has been undertaken by private, governmental and non-profit organisations for decades now, in some form as a means by which to demonstrate accountability. Information in the reports is sometimes extracted from the financial statements, but often involve the display of physical or non-monetary information. This information can be in quantitative (numbers) or qualitative (words) format, and can include both positive and negative information (Table 14.2). TBL has often been perceived as more of a hindrance than a help given the time and effort needed to collate and display the information in an appropriate format. Furthermore, there is no accounting standard which mandates the display of such information, and therefore organisations face a choice of whether to present this voluntary information or not. However increasing stakeholder pressure

Table 14.2   Triple Bottom Line reporting disclosures

Category	Positive examples	Negative examples
Financial	Profit, debt repayment, dividends paid, government taxes, conventional expenses.	Loss, interest expenses on debt, political donations.
Social	Corporate Philanthropy, health and safety programs, training programs for disadvantaged employees, gender and ethnic diversity.	Workplace accidents and fatalities, sick leave taken, industrial action, cost of regional plant closures, child labour in foreign or domestic facilities.
Environmental	Tree planting programs, voluntary environmental clean-up campaigns, recycling initiatives, environmentally efficient capital equipment, and energy consumption strategies.	Carbon dioxide emissions levels, greenhouse gas emissions, ozone depleting substance emissions, details on levels and locations of water effluent

both from the community and industry peers has placed TBL firmly on the corporate agenda. High profile corporate collapses both in Australia and overseas has resulted in 'corporate governance' being an increasingly important topic for boards to deal with. Awards ceremonies such as the annual Australasian Reporting Awards (ARA, 2004) reward both the quality of mandatory and voluntary reporting and are voted on by industry professionals.

Given the voluntary nature of TBL reports, organisations can place emphasis on different aspects. Some organisations include more financial information than others. Emphasis can be placed on colour codes to demonstrate whether organisations have met environmental targets, or information can be displayed via pie, line or bar charts. Comparative information can be presented to show variances in performance from year to year. Some organisations even have 10-year forward targets for desired emissions levels.

Table 14.3 provides an example of a Triple Bottom Line Report purely in financial format. Whilst an extensive TBL report is beyond the scope of this chapter, readers are encouraged to consult widely on other forms and structures which could include estimated pollution costs, physical data, and actual versus budget comparisons.

Another program seeking to increase the level of transparency by organisations is the introduction of the Global Reporting Initiative (2004). Begun

Table 14.3   Example of a Triple Bottom Line report along financial lines

ABC Corporation

Triple-Bottom Line operating statement for the year ending June 30, 2004

I. SOCIAL PERFORMANCE

A. Improvements $
1. Training program for handicapped workers 60,000
2. Contribution to educational institutions 100,000
3. Building upgrades to accommodate handicapped workers 65,000
4. Day Care Centre expenses for Employee's children 50,000
   TOTAL IMPROVEMENTS 275,000

B. Less Detriments
1. Postponing installing new safety devices on cutting 120,000
   machines (cost of the devices)
2. Medical expenses for workplace accidents 36,000
   TOTAL DETRIMENTS 156,000

C. NET IMPROVEMENTS IN SOCIAL ACTIONS FOR THE YEAR 119,000

II. ENVIRONMENTAL PERFORMANCE

A. Improvements $
1. Reclaiming and landscaping on company property 150,000
2. Installation of pollution control devices on 30,000
   manufacturing plant
3. Detoxifying waste from product finishing process 20,000
   TOTAL IMPROVEMENTS 200,000

B. Less Detriments
1. Cost that would have been incurred to relandscape 80,000
   strip-mining site used this year.
2. Estimated costs to have installed purification process 100,000
   to completely neutralize liquid gases
3. Fines for Breaches of environmental legislation 60,000
   TOTAL DETRIMENTS 240,000

C. NET DEFICIT IN ENVIRONMENTAL ACTIONS FOR THE YEAR (40,000)

III. FINANCIAL PERFORMANCE

A. Improvements $
1. Net Profit 200,000
2. Dividends Paid 100,000
3. Government Taxes 90,000
   TOTAL IMPROVEMENTS 390,000

B. Less Detriments
1. Interest Expense no Debt 30,000
2. Political Donations 20,000
   TOTAL DETRIMENTS 50,000

Table 14.3   Example of a Triple Bottom Line report along financial lines – *continued*

C. NET IMPROVEMENTS IN FINANCIAL ACTIONS FOR THE YEAR	340,000
TOTAL SOCIO-ECONOMIC BENEFIT FOR THE YEAR	419,000
Add OPENING BALANCE OF TBL IMPROVEMENTS AS OF JULY 1, 2003	179,000
CLOSING BALANCE OF TBL IMPROVEMENTS AS OF JUNE 30, 2004	598,000

Source: Adapted from Linowes (1972).

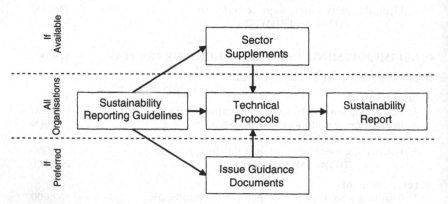

Figure 14.3   GRI family of documents
Source: Global Reporting Initiative (2004).

in 1997 as a Centre of the United Nations Environment Programme, its purpose is to develop and disseminate globally applicable 'Sustainability Reporting Guidelines'. These guidelines include social and environmental indicators which are applied using technical protocols to ensure consistency with respect to definitions, procedures, formulae and references. These are supported by issue guidance documents and specific sector supplement to ensure flexibility (Figure 14.3). The guidelines are voluntary, and like TBL Reporting, report on the organisations economic, environmental, and social dimensions of their activities, products, and services.

Mandatory environmental reporting

Whilst much of the reporting on social and environmental activities has to date been voluntary, there has been progress toward mandatory reporting.

Denmark, Norway, Sweden and The Netherlands implemented mandatory reporting requirements in the late 1990s.

In the United Kingdom, a 2004 draft of government regulations has called for mandatory reporting, termed 'Operating and Financial Review' reports, on environmental and social issues beginning in fiscal year 2005 for all 1,290 quoted companies in the UK. However the estimated cost per report is £29,000. The European Parliament's Employment Committee and The United States have also considered mandatory environmental reporting, although the major concern is that the requirements would be too costly and onerous on organisations. The Japanese Government has released a guidance document for publishing environmental reporting.

In Australia, sec 299(1) (f) of the *Corporations Act* (effective from 1 July 1998) required that 'where an entity's operations are subject to a particular and significant environmental regulation, the Director's report must include details of the entity's environmental performance'. The section was applicable to all listed and 'reporting' entities. However this provision is currently in the process of being repealed, given a Federal Parliamentary Joint Statutory Committee recommendation in 2002 that the Corporations Law was not an appropriate instrument for requiring environmental reporting, and that the section was vague and uncertain as to what constituted 'particular and significant environmental regulation'.

The increasing pressure by public interest groups and investors for social and environmental information in light of corporate misbehaviour may very well lead to mandatory environmental reporting across the board in the not too distant future. By having such systems in place now, organisations can reduce the costs of compliance if and when mandatory requirements arise.

Arguments for mandatory reporting are that it would:
- establish a minimum standard for business, which would require disclosure of both positive and negative information thereby reducing the possibility that the information is purely for public relations purposes;
- enshrine a stakeholder perspective toward business; and
- prevent a free rider problem whereby only some companies disclosed information to the market.

Arguments against mandatory reporting are that:
- the fiduciary duty of the manager is owed to the shareholders' not 'stakeholders', and as such social and environmental reporting confuses managerial obligations;
- organisations are of different size and industry makeup, therefore requiring all entities to produce the same information is too costly; and
- voluntary reporting ensures that only information demanded by the market is supplied by the market, thereby reducing information asymmetry.

14.6   Conclusion

This chapter has sought to highlight some of the more prominent aspects of the managerial and accounting based aspects of environmental management. We have seen that over the years there has been a broadening of the social contract that an organisation has with society, which has led to the proliferation in corporate social responsibility activity. We have explored briefly the different types of environmental accounting, and looked at some of the problems associated with the recognition and measurement of the environment in financial statements. Triple-Bottom Line Reporting was seen as the primary mechanism by which organisations can fulfil their social contract. The case study on Earth Sanctuaries Ltd highlights how valuation issues can complicate the measurement and recording process.

Whilst not a fully developed system, work to date on environmental accounting and reporting has enhanced awareness about the limitations of the existing financial accounting and reporting system which provides a basis by which to improve the efficiency and effectiveness of existing organisational systems. Organisations, through standards such as ISO 14000 have a greater ability to be able to track environmental costs that may have been previously obscured in overhead accounts or otherwise overlooked. This in turn leads to a better understanding of the environmental costs and performance of processes/products for more accurate costing/ pricing of products. From an investment standpoint, environmental accounting and reporting has helped to improve an organisation's investment analysis and appraisal process to include potential environmental impacts, which can lower costs to the firm. Furthermore, it supports the development and operation of an overall environmental management system. Given the increasing complexities of global business in the 21st century, and the increasing pressure placed on the natural environment to fuel development, an understanding of mechanisms which lead to more accurate and informed organisational decision making not only provide transparency, but also reduce long term costs associated with business operations, which can only lead to benefits for both business and society.

Case study: Earth Sanctuaries Ltd.

Earth Sanctuaries Ltd (2004) is an excellent case study of environmental accounting in action. It highlights the problems associated with seeking to measure the value of native flora and fauna. Earth Sanctuaries Ltd (ESL) in 1993 became the only public conservation company in Australia to operate sanctuaries for local wildlife. It financed its activities with shareholder capital issues. ESL became a listed company in 2001. ESL's sanctuaries are acquired, fenced off from predators, restocked with native marsupials, which have previously been decimated, and eco-tourism is encouraged as a revenue earning activity. Management of the sanctuaries is also an important activity with the parent company gaining a reward for providing management services to new and established sanctuaries. Tables 14.4 and 14.5 are extractions from the financial statements.

What is interesting to note is that for the 4 years up to and including the 1998 financial year, ESL produced 2 sets of accounts alongside each other. The 'financial' accounts were produced according to conventional accounting standards, whilst the supplementary 'economic' accounts were voluntarily produced by the directors so as to include a value for 'wildlife' and 'habitat', that were not legally allowed to be recorded on the 'financial'

Table 14.4   Non-current asset valuations 1995–1998 in dollars

Economic accounts	1995	1996	1997	1998
Vegetation	14,875,284	15,369,551	16,194,198	16,562,021
Wildlife	12,155,745	17,515,543	19,522,999	31,084,055
Habitat	4,806,315	17,515,543	19,522,999	31,084,055
Property and plant & other	6,749,825	7,026,059	8,609,294	12,008,925
Intangibles	927,912	688,559	1,940,648	2,616,851
Total non-current assets in *economic* accounts	39,515,081	58,115,255	65,790,138	93,355,907
Economic operating profit before income tax	31,759,685	17,685,880	6,190,537	24,462,129
Financial accounts	**1995**	**1996**	**1997**	**1998**
Property and plant & other	6,155,307	7,026,059	8,609,294	11,993,883
Intangibles	158,309	245,548	275,414	126,049
Total non-current assets in *financial* accounts	6,313,616	7,271,607	8,884,708	12,119,932
Operating profit before tax	83,512	60,662	105,518	146,622
Equity issued and paid up (including premiums)	4,739,888	6,153,597	8,914,285	10,588,180

Source: Extracted from Earth Sanctuaries Ltd annual reports for 1995, 1996, 1997, 1998 (2004).

Table 14.5   Non-current asset valuations 1999–2001 in dollars

Financial accounts	1999	2000	2001
Australian fauna	3,845,000	5,905,000	5,412,000
Property and plant & other	8,836,742	16,560,492	17,548,833
Intangibles	7,008,386	10,143,640	–
Other assets	37,462	58,801	–
Total non-current assets in *financial* accounts	19,727,590	32,667,933	22,960,833
Operating profit before income tax	1,238,009	2,055,229	(13,648,188)

Source: Extracted from Earth Sanctuaries Ltd annual reports 1999, 2000, 2001 (2004).

accounts as they could not be sold, but which the directors felt embodied value for the company due to their attraction for eco-tourism.

The company used a percentage of Australian national tourism figures as a basis to value both 'wildlife' and 'habitat', which placed a large value on the 'economic' accounts over the 4-year period compared to the financial accounts. Controversy arose as to the reliability of using this method as a basis for valuation, given the large differential in asset values between the two balance sheets, and given that no justification was provided for using national tourism figures as a basis for valuation, or the selection of the percentage rate to apply (3%) to capitalisation.

August 1998 saw the introduction of accounting standard AASB 1037 – Self-Generating and Regenerating Assets, which permitted a value for wildlife to be legally included on the accounts. As a consequence, Earth Sanctuaries in the financial year ending 30 June 1999, ceased production of its economic accounts. It had instead, included a single figure of $3,845,000 for 'Australian Fauna' in its balance sheet to represent wildlife – a marked difference from the $31,084,055 placed on wildlife in its 1998 economic accounts (Table 14.5). For several years ESL included economic values of wildlife in its annual report as a renegade supplement to conventional financial accounting information. The figures were regarded with suspicion because they used surrogate market valuation techniques. In 1999, under AASB 1037, ESL has recognised a value for wildlife in its conventional financial accounts, where it would have previously been prohibited. The new accounting standard permitted the use of cost as a surrogate for market values where no external market exists. However, the figure chosen was much more conservative than had been used previously. Whilst Earth Sanctuaries should be credited for seeking balance sheet 'relevance' by placing a value on the accounts for natural assets, questions arise as to the 'reliability' of the method chosen (Burritt and Cummings, 2002).

14.7 Questions

- Discuss the viewpoint that managerial concern for 'stakeholders' would breach the fiduciary duty owed to 'shareholders'?
- Obtain an annual report of a large organisation (can be online). Has the organisation prepared a 'Triple Bottom Line' report for stakeholders in addition to its regular financial statements? This report may be for listed companies, non-profit or government organisations. Discuss the type of information displayed in this TBL report and to what extent does it measure items in both physical and monetary formats?
- Referring to the Earth Sanctuaries case study:
  a. Was Earth Sanctuaries justified in preparing these alternative 'economic accounts' alongside conventional 'financial' accounts? What was the company trying to achieve?
  b. Do the figures in the 'economic accounts' faithfully reflect the proper value for these native assets?
  c. Should endangered wildlife be allowed to be actively traded on an open market, and if so, would it protect endangered species and increase their numbers?

14.8 Further reading

Text: Schaltegger and Burnitt (2000), Schaltegger et al (2003), Simon and Proops (2000), Pedersen (1999).

Web sites: http://www.Sustainability-Reports.com; http://www.sustainability.com; http://www.enviroreporting.comis; http://www.environment-marketplace.com/;http://www.pwc.com/gx/eng/ins-sol/survey-rep/ceo6/index.html.

15

Corporate Environmental Information

Robert Staib

15.1 Introduction

This chapter discusses the different types of information (or data) that business organisations need, to help them to achieve good environmental outcomes. It provides a hierarchical framework in which to place the information, discusses how the information can be developed and measured and the various categories of data e.g. physical data and organisational management information. Some of the issues associated with the management of the information are discussed. Many of the processes discussed in the different chapters of this book require the use of environmental information in various forms for planning, controlling, managing and reporting. Table 15.1 outlines some of these uses.

Table 15.1 Processes using environmental information

No.	Chapter name	Some uses of Environmental information
11	Corporate environmental strategy	Developing corporate strategies
12	Environmental management systems	Measuring EMS performance
13	Green marketing	Supporting green product claims
14	Financial management and accounting	Managing financial performance
16	Social and environmental reporting	Measuring and reporting organisational performance
17–18	Organisational structure and roles Changing corporate culture	Leading people and implementing organisational change
19	Interaction with external stakeholders	Interacting with and providing information to stakeholders
20–26	Environmental management techniques and tools	Using environmental assessment techniques and tools to analyse data

15.2 Measuring corporate environmental performance

Many management writers say that *what isn't measured isn't managed*. With better environmental organisational performance being expected both legally and voluntarily from organisations there is an increasing need to manage the environmental aspects of their operations and consequently this involves measuring environmental performance. There are forces both internal and external driving this need for measurement e.g. external for legal compliance, reporting and marketing and internal for process and management control.

The types of data range from physical and scientific data to management information. The data needs to be measured, collected, analysed and assessed and finally the results used for management control or for reporting to internal and external stakeholders who could include employees, management, Boards, government, community, customers and shareholders.

The environmental standards ISO 14031 and ISO 14032 (ISO, 1999a; ISO, 1999b) provide guidance in establishing a measurement framework and in linking this measurement to environmental performance and reporting. The Global Reporting Initiative (2002) provides more guidance, and links environmental measurement and reporting with the other parts of sustainability performance: economic and social. There are many ways of categorising environmental information and Figure 15.1 shows a framework that can be used to encapsulate most of the environmental measurement aspects of an organisation and which can be the basis for establishing a hierarchy of performance indicators.

15.3 Physical data collection

Many types of physical parameters may need to be measured and analysed and related to an organisation's environmental performance: environmental emissions from their processes to air, land and water; consumption of environmental resources of water, materials, energy; waste generation, disposal and recycling; impact of their operations on flora/fauna and biodiversity; and internal process variables associated with the supply chain, manufacturing/construction, delivery and product disposal (and increasingly product stewardship). These mainly appear in the boxes marked Inputs, Outputs and External Areas in Figure 15.1.

Some of this data measurement may be *mandatory* because of requirements of legislation, operating licence conditions and release reporting requirements (e.g. the Toxics Release Inventory requirements of the American EPA and the National Pollution Inventory of Australia), corporations law and certain stakeholders e.g. lenders. Some of this data may be optional or *voluntary* but will help the organisation in meeting its industry

154

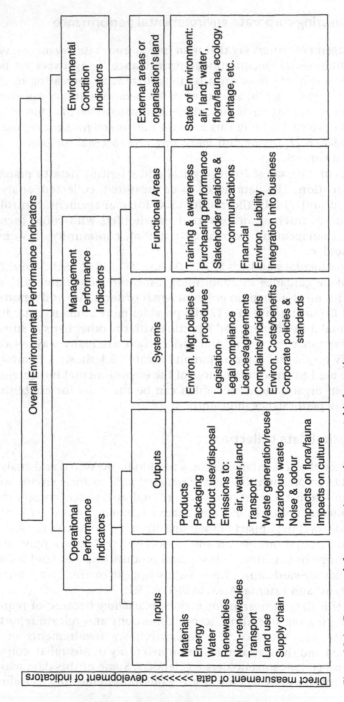

Figure 15.1 Framework for processing environmental information

Source: Based on Environment Australia (2000, p. 16); Welford, (1998, p. 170); EPA NSW (1997).

or international agreements, supporting its environmental policy and management system and assisting it to communicate with a wide range of stakeholders including its suppliers and customers.

Most of this physical data is measured quantitatively and analysed with formal scientific and engineering processes though scientific, engineering and management judgement is often needed in its interpretation and its application to the environmental performance on an organisation.

The techniques and tools described in Part IV of this book rely on the accurate and timely measurement of physical data e.g. life cycle assessment (Chapter 24) and environmental impact assessment (Chapter 25).

Over time some of this data can be collected automatically through process monitoring and control systems (e.g. effluent emissions) or regularly through operational procedures (e.g. reports from material suppliers or recyclers, invoices from electricity and water suppliers) though initially many may be measured manually or in some cases estimated. Some may require external specialist assistance to generate e.g. process and system audits, ecological assessments (Horwood, 2003).

15.4 Management information

This type of information appears mainly in the boxes marked Systems and Functional Areas in Figure 15.1. Some should be relatively easy to collect with small changes to procedures because it would already exist within an organisation e.g. number of fines or breaches of legislation. Other information may be difficult to measure or extract and will require a major change to an existing management system or process e.g. environmental costs may be hidden in general overheads (Chapter 14).

If an Environmental Management System (Chapter 12) is in place in an organisation, some of the issues of collection and assessment of management information will have already been addressed.

Often with management processes or outputs there is no way to directly measure the relevant parameters quantitatively and a qualitative measure needs to be established or an indicator needs to be developed e.g. stakeholder relations.

15.5 Developing environmental performance indicators

For an organisation to manage and evaluate its environmental performance both internally and externally it will need to develop environmental performance indicators. Each organisation will be different but a body of information on types of indicators is being developed through industry associations, corporate organisational reporting and industry research. General references to help organisations decide on the types of indicators required include the environmental standards (ISO, 1999a; ISO, 1999b) and the Global Reporting Initiative (2002).

Indicators can be used to summarise large amounts of information and:

- allow comparisons of environmental performance and trends over time;
- highlight opportunities for optimisation;
- allow the setting and pursuit of environmental targets;
- allow comparisons (benchmarking) between like organisations;
- provide feedback to the workforce on their environmental achievements; and
- facilitate communication and reporting (Jasch, 2000, p. 80).

Indicators should be: understandable to the users; comparable over time; consistently and accurately calculated; closely related to the parameters they are purporting to measure; and regularly reviewed to maintain their relevance. Where integrated or combined indicators are used the weighting assumptions should be consistent and clearly outlined. The Ecological Footprint is a summary indicator that has been applied at a regional or country scale (See Figure 2.6 and Table 10.4) but is capable of being used at an organisational scale to measure the consumption and efficient use of resources and make comparisons between similar organisations particularly in the same industry e.g. electricity generation (Barrett and Scott, 2001).

15.6 Management of data

Management of environmental data and information is an ongoing and changing process. The need and type of data needs to be regularly questioned. Different data is needed for operation control, management and reporting. Table 15.2 provides a number of steps in the general management of environmental data.

15.7 Summary

In this chapter we have discussed some of the different types of information (or data) that business organisations need to help them to achieve good environmental outcomes. There are many detailed reference sources to assist organisations to establish systems to manage environmental data and some of these have been listed. We have provided a hierarchical framework in which to place the information and discussed how the information can be developed and measured and the various categories of data e.g. physical data and organisational management information. Some of the issues associated with the management of the

Table 15.2   Environmental information management

Activity	Issues
Establish why the data or information is needed	Mandatory or voluntary, for operational control, management, environment condition or reporting
Identify current sources of data and shortfalls	In-house, industry associations, suppliers, literature
Develop data management plan	Measurement objectives, environmental values to be protected, organisational parameters to be measured, indicators to be developed, resources required (equipment, people, systems), procedures and responsibilities
Establish data development programs	Ongoing process to enhance and improve data collection, analysis and reporting
Develop data specifications for measuring, collection, collating, analysing and communicating	• Unit, parameter, indicator and their definitions • Methods of measurement, calculation, collection, verification, storage • Assessment processes • Measurement period • Form of trend analysis • Short, long, medium term targets for data • Relationship to the operational and management aspects of the organisation
Collect and use data to management environmental performance	• Does it show the organisation is improving its environmental performance positively and then significantly? • Use combined indicators e.g. to measure sustainability • Compare with other organisations' or industries' norms
Show legal compliance	At the most basic level an organisation's environmental data should show it is complying with all its mandatory requirements
Review data plan, enhance and revise	Periodic review of the adequacy of the data/indicator

Developed from: the author's experience; Horwood (2003); Environment Australia (2000); EPA NSW (1997).

information are discussed. Other chapters of the book discuss different ways in which information in used for planning, controlling, managing and reporting. Finally in Table 15.2 we outline a simple process to establish and manage environmental data to enable an organisation to improve its environmental performance.

15.8   Questions

* Choose an industry and a number of organisations within that industry (e.g. automotive, electricity, white goods) and using published environmental data calculate and compare the eco-efficiency of their products or services for a number of indicators e.g. energy used in the manufacture per product, toxic waste produced per product.
* Using published environmental data for an organisation over 5 years (at least), identify whether it is reducing its impact on the environment per product and overall. Do its reporting and marketing claims match its published data?

16
Social and Environmental Reporting
James Guthrie and Carol Adams

16.1 Introduction

This chapter outlines a number of issues associated with Corporate *Social and Environmental Measurement and Reporting* (SEMR). It explores several external drivers and a number of internal drivers for SEMR. A recent survey of Australian SEMR is reported and international guidelines for measurement and reporting are detailed. A number of benefits from corporate social and environmental measurement and reporting are outlined. The final section explores several future directions for SEMR and why it is important to environmental management in an organisation.

It has been widely acknowledged that the impact of business organisations on society is not only economic, but also extends to environmental and social outcomes. Therefore, to focus solely on economic performance is considered too narrow – societal expectations are no longer confined to profit generation and the provision of goods and services. There has been increasing demand for the benchmarking of company performance against non-financial aspects, such as environmental and social performance (e.g. Estes, 1976; Guthrie and Matthews, 1985). From the perspective of the Corporate Social Responsibility (CSR) research, the traditional financial accounting framework is too narrow (Guthrie and Parker, 1990). Alternative measurement and reporting methods that have been proposed in the CSR research include social accounting and environmental accounting. The management and reporting of social and environmental performance has attracted greater attention since the report *Our Common Future* (World Commission on Environment and Development, 1987) which put sustainable development firmly onto the international political agenda (Elkington, 1998, p. 55). In the past decade a plethora of environmental reporting and social accounting guidelines have been released. However, there is no universally accepted framework.

16.2   External drivers for corporate SEMR

There are two key reasons put forward for the increased emphasis on environmental and social reporting by corporations. The first is a growing recognition that corporate power entails responsibilities to various stakeholder groups. Stakeholders include shareholders, but also neighbours, workers, consumers, media and governments. Business organisations face increased attention to their activities and need to provide a moral justification for their activities to the general public. The moral justification for behaving responsibly is often translated to a business reason for change (Adams and Harte, 1998).

The second is a recognition that it is sensible business practice to behave responsibly and build trust with stakeholders. Such measures can improve a corporation's image and reduce the long-term financial risks associated with poor environmental or business practices.

These reasons are intertwined and individuals within organisations often give both as a reason for behaving responsibly (Adams, 2002). Clearly, a range of stakeholders has legitimate concerns with the operation of a business and these concerns help to drive forward the processes of social and environmental reporting.

In examining external drivers, internationally, there are a number of nations that stand out. Some, for example Scandinavian countries and the Netherlands, have mandatory corporate environmental performance reporting requirements (KPMG, 1999). In the US companies have to submit emissions data to the Environment Protection Authority, which is made available publicly. Also the US Securities and Exchange Commission, Canada's Securities Commission and the UK *Companies Act* require the disclosure of social and environmental information which affects current or future financial performance (KPMG, 1997). The EC Green Paper, *Promoting a European Framework for Corporate Social Responsibility*, is a recent European initiative aimed at improving corporate social and environmental disclosure (Simms, 2002).

The contemporary requirements of pension funds and investment products creates external pressure. For instance, in the UK, pension funds are required to incorporate their socially responsible investment policy in their statement of investment principles making it easier for investors to apply pressure to companies (Friedman and Miles, 2001; Olivier, 2001; Skorecki, 2001b). For instance, Morley Fund Management, which manages assets equivalent to 2.5% of the UK stock market, excludes any FTSE-100 company without an environmental report (Dickson, 2001; Parker, 2001; Skorecki, 2001a). In Australia, the *Financial Services Reform Act* requires sellers of investment products to disclose the extent to which labour standards, and environmental, social and/or ethical considerations are included in their investment decisions (Guthrie and Carlin, 2004; Haigh, 2004; Macken, 2002).

Another external pressure is from Non Government Organisations (NGOs), which are increasingly willing to buy shares and take their corporate battles to the annual general meeting. For instance, Greenpeace used BP's 2001 annual general meeting as a forum for climate change issues and to raise awareness of the impacts of Petro China's Tibetan pipeline (Buchan, 2001; Frey, 2002). In another instance, the Australian Wilderness Society forced the Commonwealth Bank to table a motion to ban the bank from investing in Tasmanian logging operations (Gettler, 2002; Potts, 2002). In the US, student criticism of Nike for alleged labour rights abuses at its Indonesian and Mexican factories led to a threatened boycott of Nike university logo clothing. As a result of this activism, Nike hired an independent agency to monitor employment practices overseas (Edwards, 2002).

16.3 Internal drivers for SEMR

Businesses now acknowledge that improved stakeholder consultation is in their interest (Kok et al, 2001). However, companies may attempt to alter perceptions of their performance through reporting rather than through social and environmental measurement and performance improvement (Fowler et al, 1999; Hooghiemstra, 2000).

Adams (2002) reports the responses of people responsible for environmental reporting in British and German chemical companies and uncovers an ambiguous set of motivations including: public and political pressure; an internal desire to act responsibly; and the desire to improve corporate image.

Another internal pressure is the management desire to manage risks and potential liability. For instance, a fatal gas explosion at Esso's Longford plant in Victoria, Australia in 1998 demonstrated the potential for corporate liability. The coroner's verdict concluded that Esso was solely responsible for the accident, having failed to conduct detailed periodic risk assessments and a more comprehensive hazard audit in line with its own policies (Gale, 2002; Kletz, 2001; Lapthorne, 2002; Madden, 2002; Pheasant, 1998).

16.4 Voluntary reporting and GRI guidelines

Recently, the annual reports of the Australian top 50 Australian Stock Exchange (ASX) listed companies were examined, using an Extended Performance Reporting Framework (Yongvanich and Guthrie, 2004) to analyse Corporate Social and Environmental Reporting (CSER). There were 49 companies in the final sample. The results are shown in Table 16.1. They demonstrate a lack of consistency in social and environmental reporting amongst the Australian sample.

Table 16.1 Frequency of social reporting by the top 50 ASX listed companies

Social Reporting Categories	No *	Social Reporting Categories	No *
1 Labour Standards			
Organisational culture	47		
Rewards, performance measurement system and alignment	37	**4 Ethical considerations** Human rights	
Training and education	40	– Strategy and management	5
Employment	33	– Non-discrimination	12
Health and safety	38	– Freedom of association and collective bargaining	3
Management/labour relations	27	– Child labour	1
Diversity and opportunity	23	– Forced and compulsory labour	1
2 Environmental considerations		– Disciplinary practices	1
Materials	18	– Security practices	0
Energy	34	– Indigenous rights	14
Water	19	Society	
Biodiversity	9	– Bribery and corruption	7
Emissions, effluents, and waste	35	– Political contributions	16
Suppliers	8	– Competition and pricing	15
Products and services	22	Product Responsibility	
Compliance	40	– Products and services	7
Transport	4	– Advertising	0
Overall	1	– Respect for privacy	17
3 Social considerations			
Community	43	**TOTAL NUMBER OF COMPANIES**	49
Customer health and safety	22		

Source: Yongvanich and Guthrie (2004).
Note: No = number of companies reporting on a category.

A range of tools and guidelines for social and environmental reporting are available. Some of these are: Global Reporting Initiative Sustainability Guidelines (GRI, 2002), Public Environmental Reporting: an Australian approach (Environment Australia, 2000); AA1000 Framework (ISEA, 1999); and ASIC Section 1013DA Disclosure Guidelines (ASIC, 2003).

The international environment standard ISO 14031 provides a framework for environmental measurement and reporting and is discussed in Chapter 15. The Global Reporting Initiative extends this to link with the social and economic.

16.5 Benefits from corporate SEMR

This section examines a number of benefits to companies which demonstrate social responsibility and account for their social and environmental impacts. One benefit is the better recruitment and retention of employees. Another is improvement in internal decision making and cost savings. Further benefits are an improved corporate image and relations with stakeholders and the possibility of improved financial returns. Not all of these benefits are easily quantifiable!

Better recruitment and retention of employees

Acting responsibly and being accountable for social and environmental impacts assists organisations in attracting and retaining talented staff (Adams, 2002; Bernhut and Bansal, 2002; Simms, 2002). An international survey conducted by Hill & Knowlton found that 88% of British businesses believe that social responsibility will become increasingly important in employee recruitment and retention (Hill & Knowlton, 2002; Simms, 2002).

Improved internal decision making and cost-savings

Organisations producing social and environmental reports develop better internal control systems leading to better decision making, cost savings and continuous improvement (Adams, 2002). Improved operational and process efficiency results in reduced risks and safer workplaces (King, 2002; Simms, 2002). Open communication of good *and* bad news to employees and stakeholders via socially responsible reporting facilitates these improvements.

Improved corporate image and relations with stakeholders

Improved disclosure of relevant social and environmental issues by companies contributes to minimising the risk of consumer boycotts (Adams, 2002). Better communication with stakeholders and the broader community (Anand, 2002; Bernhut and Bansal, 2002; Marx, 1992/93) improves their understanding and reduces the level of criticism, leading to reputation enhancement and competitive advantage (Adams, 2002; King, 2002).

Improved financial returns

Socially Responsible Investment (SRI) involves investors taking personal values and social concerns into account when making investment decisions. Ethical investment funds consider investors' ethical values and financial needs and the social impact of the companies they invest in. Friends Provident set up the first ethical fund in the UK in 1984 (Brown, 2003) – there are now approximately 55. This number is forecast to increase.

In Australia SRI managed funds grew 32% to AUD13.9 billion in the financial year 2001–2002 (Gibson and Kendall, 2003) while assets of managed funds as a whole declined. Joyner and Payne (2002) identified a positive link between a firm's value, business ethics, corporate social responsibility and financial performance, indicating that investors may be profiting from SRI while adhering to their values.

The difficulties of providing an accurate cost benefit analysis of CSR initiatives (Adams, 2002; Bernhut and Bansal, 2002; Evans, 2003) lead to challenges for managers trying to convince company directors of the advantage or usefulness of good reporting and disclosure systems (Simms, 2002). An interviewee in Adams' study said:

> We strongly believe it is good for us to report on environmental issues, but we can't measure the effects in financial terms. Our CEO ... likes things to be measured and quantified, but you can't put a figure on the impacts on your image if you had not passed the EMAS certification (Adams, 1999, p. 41).

An extended reporting framework

There are many examples of how to undertake SEMR. We provide only one for illustration purposes. Yongvanich and Guthrie (2004) briefly reviewed three reporting approaches, namely the intellectual capital, balanced scorecard and social and environmental reporting. Yongvanich and Guthrie demonstrated that the theoretical emphases of these reporting approaches are complementary and should be integrated and developed into an extended performance reporting framework (EPRF). The EPRF provides both economic and non-economic performance information and is expected to provide a more complete account of company value and performance.

16.6 Summary and conclusions

Corporate Social and Environmental Measurement and Reporting (SEMR) has been growing since the early 1970s with major companies now externally reporting because of legal regulation or because they see some advantage in undertaking this activity.

This chapter has outlined some of the external and internal drivers for SEMR. It has provided some contemporary evidence of reporting practices and highlighted the potential benefits to corporations in improving their environmental and social management, measurement and reporting, which can ultimately reduce their negative impact on the environment and society.

Case study: Camelot

The results of a survey of 6,000 people from 8 stakeholder groups by Camelot, the UK's National Lottery, prompted this comment from Sue Slipman, social responsibility director:

...stakeholders do not always share common interests. Sometimes the correct response will be for the management to reject the calls of stakeholders. But social reporting remains valuable, because it provides an informed basis for explaining the company's actions (Pike, 2000, p. 18).

Advantages of reporting and disclosure were also outlined by the interviewees in Adams' (1999; 2002) studies: 'There is a very important internal message about the performance of the company and where we are doing well and where we are not. So that when that information goes out, the businesses can then benchmark their own performance against that of the Group. And there are some businesses where performance is sadly lacking. So they have got an opportunity with that report to help boost their performances within their own businesses because they have actually got a measurement that they can compare with. So it is a very important internal management tool' (Adams, 2002, p. 236).

'I think one of the big advantages is credibility with your external audience ... you are saying you have got nothing to hide, you are explaining where there are problems as there inevitably are. You are explaining why they have happened. Obviously you are setting yourself targets ... within the organisation you need to be measuring these things ... if you don't measure it, you can't manage it' (Adams, 1999, p. 42).

16.7   Questions

- As public and private companies use resources and create waste and pollution, should they be required to undertake annual external reporting of their environmental and social performance and make the results public?
- Should an organisation that makes social claims be subject to external questioning at its annual general meeting and be asked to verify its claims?
- Choose one annual report of a top 50 ASX listed company in your country and, using an Extended Performance Reporting Framework (or some other form of social and environmental framework), identify the items reported and the measurements used in the external reporting.

Part III
Culture and People

17
Organisational Structures and Roles

Andrew Griffiths, Suzanne Benn and Dexter Dunphy

17.1 Introduction

This chapter looks at the impact organisational structures have on the pursuit of sustainability, offers some state of the art alternatives to restructuring organisations and finally outlines some change agent roles and responsibilities for pursuing sustainability within organisations.

Organisations around the world have been subjected to increasing pressure to improve their environmental performance. Much of the current research and debate has focused on practices and strategies that organisations and managers adopt to deal with ecological concerns (Hart, 1997; Hutchinson, 1996). However, there is a growing awareness about which structures assist organisations to enact these strategies (Hoffman, 2000), but there is a need (Starik, 1995b) for more research to examine the relationships between corporate structures and ecologically sustainable organisations (ESOs).

17.2 Impediments to sustainable structures

There are three ways in which current corporate structures impede sustainability (Sharma and Vredenburg, 1998; Starik and Rands, 1995c). *Firstly*, current corporate structures insulate organisation systems and processes from a broad range of environmental information (Griffiths and Petrick, 2001). This is because they have few structural elements, such as specialised departments that have the expert knowledge to recognise, act on and transfer to other parts of the organisation information of an environmentally relevant nature. Such organisations typically have a reactive or minimal compliance stance to the adoption of environmental management strategies (Hunt and Auster, 1990; Roome, 1992).

Secondly, the established routines and organisational systems of many corporations seek to protect and promote the status quo. Schon (1971) refers to this as the dynamic conservatism of social systems. As a result, new practices and theories are defined as threats to established ways of working. In particular, the corporate innovations which support sustainability are a threat to conventional command and control style management systems and often require new structures, strategies and organisational capabilities. Managers who have risen to power in command and control systems frequently resist the transition to alternative structures.

Finally, current structures limit or deny access to a range of stakeholders whose participation is vital for the pursuit of sustainability agendas. Traditional organisations tend to be focused on a limited set of stakeholders, in particular, boards of directors and shareholders (Hart, 1995). Sharma and Vredenburg (1998) argue that such traditional or reactive companies have not developed capabilities for stakeholder integration. The initiatives of other stakeholders – particularly the workforce, unions and 'greenies' – are seen as hostile, disruptive forces that may harm the basis of current corporate performance. In response to initiatives from these stakeholders, some corporations adopt aggressive and defensive practices which undermine the development of the broader stakeholder accountability and involvement needed to support moves to create more sustainable practices (Beder, 1997).

These three limitations refer primarily to large traditional corporations that are hierarchical in design, have developed systems of managerial practice based on a strong centralised command and control style of management, and pursue reactive approaches to ecological and human sustainability issues (Hunt and Auster, 1990). Despite these limitations, an active search for new organisational forms is taking place (Griffiths and Petrick, 2001).

17.3 Characteristics of ecologically sustainable corporations

Conditions that characterise ecologically sustainable organisations (ESOs) (Korten, 1995; Weizsacker et al, 1998; Hunt and Auster, 1990) include:

- smaller corporate entities and structures will be more responsive to environmental concerns and less powerful and less inclined to dismiss government and societal attempts to regulate them (Korten, 1995);
- limited government regulation can be used to shape corporate environmental behavior to comply with ecological standards and adopt proactive environmental management practices, such as Total Quality Environmental Management (TQEM) to reduce waste and pollution (Hoffman, 1997);
- increased power should be devolved to individuals and local communities in order to create a citizen-inspired agenda for local ecological sustainability (Korten, 1995); and
- future ESOs will play an active role in creating self sufficient communities where production and use aligns with community needs.

Shrivastava (1995) and Weizsacker et al (1998) suggest that corporations can change other organisation and industry structures by becoming much more efficient at modifying inputs, through-puts and outputs to reduce their negative impact on the environment and gain competitive advantages.

Secondly, there exists a range of corporate strategies that have consequences on the structure and content of corporate environmental programs e.g. environmentally responsible supply chain management may require little change to organisations' structure and orientation. Other strategies such as industrial ecology may require significant changes to an organisation's operating system, definitions of waste and relations to its value adding chain (Senge and Carstedt, 2001; Cohen-Rosenthal, 2000). ESOs will require internal structural changes to their operations systems, such as the implementation of TQEM programs, Environmental Management Systems (EMS) and a broader understanding of what constitutes social and environmental 'best practice' i.e. transforming structures to ESOs is both desirable and achievable.

17.4 Alternative structures for sustainability

In this section three alternative organisational structures are identified and proposed – the *network*, the *virtual organisation* and the *community of practice*. Each is an ideal type which in reality may vary and appear in hybrid form. However we identify their general characteristics and the prospects that they provide for enhancing human and ecological sustainability. These forms can be replicated at the level of groups, business units and whole organisations, making it not only possible to create new organisations based on these ideal types, but to also retro-fit existing organisational structures to capture sustainability benefits (Griffiths and Petrick, 2001).

Network organisations

The network organisation has emerged as a powerful alternative to traditional vertically integrated organisational structures (Nohria and Eccles, 1992). Network organisations may take a variety of forms but fundamentally they are organisations that give the advantages of larger size, whilst remaining small. In these organisations, the centre retains some key areas of decision making, e.g. major strategic moves, ceding considerable autonomy to smaller relatively independent units. For instance TCG, a company that operates in the technology and software support and services industry, is a small network of 24 independent firms that cluster together to provide the advantages of size while maintaining the flexibility and innovativeness of small firms (Mathews, 1992; Quinn, 1992) e.g. flatter hierarchies and minimum use of formal rules.

A major strength of *network* structure lies in its ability to grow (by adding on new firms) whilst keeping the constituent units small, flexible, responsive and innovative. Networks are sustainable primarily because of their ability to respond quickly to changing market conditions, to be flexible in meeting customer requirements and to transform markets through the speedy development of new products and services. However, networks have to be coordinated differently from traditional hierarchical organisations – they often appoint network 'brokers' who coordinate the various elements of a network value-adding chain. Networks are strongly dependent on the skills, dexterity and knowledge base of their employees – whilst traditionally associated with craft work, new network forms have emerged in the new 'craft industries' of biotechnology, software and information technology. Networks are relevant to sustainability. Firstly, networks are increasingly recognised as a major source of innovativeness in new product and service developments. In Germany and Denmark clusters of small firm networks have been leaders in green technology developments. Secondly, network structures are reliant upon free flowing information and communication – they are therefore appropriate structures for capturing and diffusing through the network information relevant to sustainability outcomes.

Virtual organisations

Another alternate organisational form that can support sustainability is the *virtual* organisation (Hedberg, 1997). The virtual corporation can be seen to be designed on several levels. At one level the virtual corporation can be interpreted as an organisation with a limited life. At first appearances, this does not appear to be a 'sustainable practice'. In these cases, the issue is not about sustaining the organisation in terms of longevity, but rather recognising that in the pursuit of sustainability there will be a need for limited term projects – organisations that will come together to solve or address important issues and disband once they have been addressed.

Alternatively, a *virtual* organisation can give the impression of size, yet be small in terms of numbers of employees i.e. corporate sustainability is not just about large corporations but can also include smaller more nimble corporations that have a virtual facade. Such organisations, as Amazon.com (an internet book retailer) and Land's End (textile and clothing manufacturer) can service via the Internet and electronic commerce and operate globally with a small number of employees with structures that are flexible and compete on speed. They will tend to leave a minimal environmental footprint, however they will need to be responsible for the environmental impacts of their suppliers and distributors.

Communities of practice (See Epistemic communities, Section 5.4)

Unlike the previous two organisational forms, *communities of practice* are not clearly defined entities. Communities of practice have amorphous and fluid structures that form around areas of interest, expertise and/or project orientation (Brown and Duguid, 1991). An example is a community of professionals who gather information, pass on knowledge and contribute to the development of their field of expertise. They generate the diffusion and acceptance of explicit and tacit knowledge that can be transferred into innovative solutions and actions within formal organisations. Increasingly, communities of practice rely upon and are assisted by cutting edge information technologies that provide the potential for innovation and the ability to generate solutions quickly. They may be difficult to design, but rather emerge organically and are continually being formed and reformed.

Whether they are located within formal organisations or arise independently, *communities of practice*: are reliant upon structures which enable them to take on new members; acquire new information and bind people to each other through common interests, desire for learning and an enhanced ability to achieve collective and individual goals; are characterised by a reliance upon both formal and informal processes for skills development and learning; and need to establish a core of people responsible for creating and sustaining the community's collective memory. Their contribution to the attainment of sustainability outcomes lies in their ability to collect, process and diffuse knowledge of a technical and specialised nature and translate this knowledge into innovative and rapid solutions (Griffiths and Petrick, 2001).

17.5 Structures and sustainability

Structures: that capture and use ecological information

An important element in the generation of sustainable corporations is the creation of organisational structures and systems that are capable of capturing, processing and making sense of environmental information. King (1995) notes that organisations with limited access to environmental information are more likely to make ecological and environmental blunders. To avoid this and capture the benefits of environmental awareness, there must be multiple and diverse information entry points into the organisation. If critical information about the environment is confined to specialist units it will fail to diffuse throughout the organisation and is unlikely to be used in strategic decision making. *Communities of practice* can help assist with the diffusion and acceptance of such vital information by creating a broad range of information entry points to reinforce the creation and generation of corporate wide environmental strategies.

The challenge is to create corporate structures that effectively internalise the environmental information needed to meet regulatory standards and corporate competitive strategies based on sustainable practices. E.g. Volvo established environmental programs by creating four corporate-wide 'working groups' focused on a broad environmental agenda which: examined structures and systems to capture the benefits of product recycling; structured organisational systems for acquiring and diffusing environmental information; and reviewed production processes for environmental efficiencies to enable it to meet European Union regulatory standards on environmental management.

Structures: that incorporate employee knowledge

Both the human and ecologically sustainable literatures emphasise the importance of employee knowledge in generating adaptive and responsive organisations. Hierarchies are killers of initiative and innovation and impede the effective utilisation of employee knowledge. Team-based organisational structures enable employees to have input, discretion over decision making and provide them with the information and resources to deliver organisational outcomes. Such strategies are long-term and for team-based organisational structures require large investments in time and energy. As a consequence, managers need to make the shift from a short-term orientation to the longer-term strategies of developing these people centred, service orientated structures (Quinn, 1992). Team-based organisation structures which include, project teams and *virtual teams*, based on new communication technologies, provide the basis for capturing and utilising the employees knowledge needed to create effective sustainable corporate strategies (See Case Study).

Structures: rapid response to sustainability opportunities and threats

The structures for achieving rapid responses to sustainability opportunities and threats can give corporations the advantage of speed by becoming 'ambidextrous'. On the one hand, corporations need to maintain the day-to-day operation activities that create valuable products and services. On the other hand, they need to support the ongoing transformation of these activities to move them in the direction of human and ecological sustainability. Some corporations have already created organisational structures that position them to take advantage of these processes by creating supportive *networks*, *virtual* organisations and *communities of practice* e.g. Scandic Hotels moved towards rapid efficiency gains through the adoption of value adding activities and flexible structures by building on the capability of its employees. Anticipated efficiency gains were shared with employees by investing in their skills with emphasis on developing and training to identify value-adding opportunities (Dunphy et al, 2003). Employees developed and used a range of metrics such as quarterly benchmarking reports and an environmental index. In its first year in its Nordic hotels reduced consumption: energy 7%, water 4% and unsorted waste 15%, with estimated benefits of US $800,000 (Nattrass and Altomare, 1999, p. 92). Through investment in such value adding activities, Scandic built on its cost approach to deliver further efficiency improvements in resource utilisation (Dunphy et al, 2003).

17.6 Change agent roles for sustainability

The various roles internal and external to organisations can contribute to sustainability initiatives. They can make a distinctive contribution to moving organisations toward sustainability; in collaboration, they can have an irresistible impact. We begin our discussion with roles internal to the organisation.

Boards of management

Directors have heavy legal responsibilities and can be liable to major penalties if the organisation is not compliant. They also have ultimate responsibility for appointing the CEO and signing off on the organisation's business strategies. Boards are increasingly concerned with issues of sustainability as more companies face heavy penalties for failure to comply to environmental and health and safety legislation, as corporate reputations are damaged by the revelation of unethical practices and as ethical investment funds press for sustainable policies as the condition for providing investment capital. They are realizing that to press for shareholder returns at the expense of the interests of other stakeholders can be costly for themselves and the company and damage their own reputations as well as the corporate reputation. Many are also realizing that sustainability represents a range of unexploited future business opportunities which can create future growth.

The chief executive officer (CEO)

The role of the CEO is vital in terms of both symbolic and practical leadership. It is no easy role. Ford Chairman William Ford was faced with a dilemma after he made a series of major sustainability initiatives, including a model citizenship report called *Connecting with Society* in which he said that the popular sports utility vehicles (SUV) were unsustainable given the petrol they consumed and that Ford aimed to be a model company for the 21st century, particularly in the area of sustainability. Shortly after there was a recall of six million defective Firestone tyres shown to have triggered many accidents, including deaths. Ford vehicles, particularly SUVs, were the major users of the tyres. Firestone had attempted to cover up the growing history of accidents and is now mired in lawsuits. Firestone was the supplier but Ford's reputation was affected as well as its profits. Addressing the 2000 Greenpeace Business Conference, Ford said: 'This terrible situation – which goes against everything I stand for – has made us more determined than ever to operate in an open, transparent and accountable manner at all times' (Elkington 2001, pp. 118–120).

The CEO of a public company faces the dilemma that many market analysts and investment funds seek short term returns and place too little emphasis on the importance of the issues discussed in this book. The CEO who is attempting to build a sustainable and sustaining corporate culture faces a daily performance evaluation by a sharemarket which traditionally places little value on this. Nevertheless many CEOs manage to generate short-term gains by 'picking low hanging fruit' while quietly investing in building the capabilities of the corporation to generate medium to long-term performance, including performance due to sustainability initiatives. Effective CEOs and executives are active in understanding the changing context in which the organisation operates and place themselves where they will be sensitised to emerging changes in society. A problem in creating significant change in organisations is that many executives move in very conservative circles and live in elite enclaves cut off from mainstream social life. When organisations actively relate to the community this offers opportunities for senior executives to break out of their circle of privilege and engage with other people.

Managers, supervisors and team leaders

Managers contribute to the strategic process and actively translate the strategies into practical action plans. If they are to be effective, these plans will include achievable but challenging sustainability goals and involve the introduction of processes and systems that embody ecological and human sustainability principles. Supervisors and team leaders are the critical front line of both incremental and transformational change. Their support and feedback is vital to a successful change program.

General employees

Employees often see themselves as having more limited power than executives and managers. They have more limited authority but general employees make or break organisational strategies. A strategy that is not translated into the day-to-day operational work of the organisation is like a bird without wings – it never gets off the ground. Senior managers are the architects of change but general employees are the builders. The intelligence, commitment and skills of general employees move an organisation forward on the path to sustainability. The moments that test whether an organisation is serious about sustainability occur as members of the workforce face customers across counters, on a telephone or on line, negotiate with suppliers, respond to new legislative requirements, dispose of waste. In the process of change, there are also opportunities to innovate, to suggest new or improved ways of doing things, to bring in knowledge acquired elsewhere. Employees have another source of power, they can vote with their feet and leave a company that is acting unethically or not taking new sustainability initiatives.

Human resource (HR) specialists

HR specialists are an integral part of the change planning process, including working with the senior executive team. They must also be active in creating and maintaining the guiding coalition that leads the change process and need to provide technical expertise on HR issues. They are an alternative channel for the expression of employee ideas and concerns as the change process evolves. Another important part of their responsibility is to ensure that the organisation's reward system does not lag behind the innovations but supports behaviour consistent with the phase of sustainability being established. Building the skill base needed to support relevant sustainability practices is their particular responsibility. This consists of the competencies needed to continually reshape the organisation for future sustainable performance – the competencies for enhancing the organisation's capacity for change. They build both technical and reshaping skills through recruitment, consultancies and education of employees.

Organisational development (OD) specialists

OD practitioners are particularly important in organisational change. Not all organisations have OD specialists or refer to them in this way. By OD specialists we mean change agents with highly specialised skills around the process of corporate change. They are professionals with training and experience in techniques such as team building, conflict resolution, counseling and intergroup relations. They are accustomed to working with the hot human process of change as it occurs and taking emotional reactions as a normal part of any significant change and being unfazed by them. They are accustomed to: working in ambiguous situations where they have little authority; collaborating with others to design the ongoing process of change; working on the edge of chaos; and keeping organisational change processes on track and productive.

Industrial engineers and environmental specialists

In manufacturing industry, technical specialists with an engineering background play key roles in planning and operations. In most cases their training does not equip them with knowledge of how to make organisational change, particularly on the human side of the process. However some bring good interpersonal skills learned in the workplace and considerable experience in making change. Where they have a grounding in Business Process Reengineering this can prove invaluable in redesigning product flows to make the flows more efficient and sustainable. Their technical knowledge needed in moving to sustainability can combine with the skills of HR and OD specialists to be a winning combination. All too often, specialists from these diverse backgrounds speak different languages, don't communicate with each other or appreciate the relevance of the other professionals' training and knowledge. This can be a significant obstacle in moving toward integration, particularly in the latter phases of sustainability where integration is vital.

17.7　External change agent roles

In the sustainability movement, external change agents have played major roles in bringing pressure to bear on organisations to adopt more sustainable strategies. This external pressure has been at times adversarial in both human and ecological areas of activity. While there will be a place for adversarial activists in the foreseeable future, a shift in emphasis has taken place as more organisations move beyond compliance to launch sustainability initiatives on a voluntary basis. New collaborative alliances are taking place across the boundaries of organisations and external change agents have important roles to play in these alliances. Consulting companies have set up specialised practices to provide advice on a variety of sustainability issues.

Investors (See Section 9.8)

Investors control the flow of capital to companies directly and through brokers and funds. There is no more powerful pressure for change than the withdrawal of investment from public companies or the flow of capital to them. It is vital therefore that investors support companies that are working to implement sustainability policies and withdraw investment from those that are not. Customers can exert similar pressures through shifting their buying to favour sustainable products and services. Consumers are increasingly demanding transparency in terms of ingredients and components in products (e.g. information on genetic modification of foods). Supporting increasing transparency makes an important contribution to advancing the sustainability movement.

Community activists

Community activists, concerned citizens, intellectuals and scientists also have important roles in demanding transparency in company operations, assembling the best available knowledge about the impact of particular kinds of products and services (e.g. chemicals), and bringing external pressure to bear on companies that avoid their responsibilities to their workforces, communities or the environment. In addition, as more companies move to an 'inclusive leadership' approach that welcomes the participation of a range of stakeholders in achieving corporate sustainability, there are emerging opportunities to collaborate on new initiatives. Community activists also have a vital role in speaking for the natural environment – they are often closest to it and best able to provide a voice to protect it.

17.8　Conclusion

While a significant amount of time and energy has been devoted to the creation and generation of environmental corporate strategies, there has been virtually no attention given to understanding the ways in which different corporate structures impede or facilitate progress toward human and ecological sustainability. We have tried to fill this gap by bringing together knowledge from several traditions. Our basic approach has been that traditional structures will not support the adoption of the changes in business practices necessary to achieve human and ecological sustainability. Corporations, large and small, will need to experiment with alternative organisational structures and systems and this is already occurring.

In this chapter we have discussed three alternative organisational structures – *networks, virtual organisations* and *communities of practices*, and the changed organisational roles needed to move organisations toward sustainability.

Many emerging organisations will never become large traditional bureaucracies. Existing bureaucratic organisations will need to adopt in part these new corporate forms or modify the rigidity of their current structures. Furthermore, the operations and production management role will need to influence the design and structure of organisations in response to these sustainability challenges. Through engagement with technologies and processes (e.g. industrial ecology, environmental management systems and total quality environmental management) operations management will play an integral role in shaping and re-designing an organisation's response to sustainability issues.

Case study: enabling structures at Fuji Xerox

The highly successful Fuji Xerox Eco Manufacturing (EM) Plant at Zetland, Sydney demonstrates that a positive relationship between human and ecological sustainability can generate transformative change. The concept of EM involves detailed analysis of "why things fail", remanufacture of products to eliminate future failures and supply to local operators with high quality locally reprocessed parts. EM takes used components and re-engineers/reassembles them into "new" products ensuring that the production process and the products do not have adverse environmental effects.

A key success factor at Fuji Xerox has been a small group of skilled, innovative and committed managers willing to listen to staff, customers and other stakeholders. Staff are assigned to teams, each team being responsible for quality, engineering and production capacity around products or product groups e.g. print cartridges or lasers. The product-based team structure promotes multi-skilling, enhances communication around problem identification and solving, builds expertise and cumulative experience and ensures that improved quality is constantly built into the work process. Managers see this as the leading cause of the high level of innovation in the plant. It has led to a close working relationship between the engineers and the production workers and joint ownership of the production targets and product quality.

The plant is systematically building the human capabilities of its staff. Staff members are offered a range of developmental opportunities and training in "people management". Employees are well remunerated and turnover has been low over the last eight years, making Fuji Xerox an employer of choice. The shift towards a more enabling and committed culture has been significant. The organisational changes illustrate the links between an organisational culture of innovation and one designed to deliver sustainability (Benn and Dunphy, 2004a).

18
Changing Corporate Culture to an Environmental Ethos

Suzanne Benn, Andrew Griffiths and Dexter Dunphy

18.1   Introduction

In this chapter we propose an integrated perspective on organisational culture designed to progress the ecological sustainability of the corporation. We indicate how to make an *incremental transition*, or in some cases, a *transformative leap* to the fully sustainable and sustaining corporation. Case study examples are given of organisations we have studied which have made both incremental and transformative shifts towards ecological sustainability as a result of cultural change.

We argue that the organisational move to *ecological sustainability* is reliant upon *human sustainability*: the development by the organisation of the human capabilities and skills that enable more consistent compliance, the implementation of eco-efficiency measures and forward planning for sustainability i.e. we argue that the concepts of human and ecological sustainability need to be merged.

This chapter builds on recent research which has indicated a relationship between corporate culture, human resource policies and the successful implementation of the Environment Management System (EMS) as a strategic business and risk management tool (Daily and Huang, 2001; Wilkinson et al, 2001). This research concluded that EMS programs are more successful if training, empowerment, teamwork, and rewards are addressed and suggested a relationship between human and ecological sustainability in the form of a phase model.

18.2   The drivers for change

An integrated perspective on corporate sustainability is necessary to capture the complex set of corporate responses to the wide array of influences impinging on the contemporary marketplace. One result of the merging of the internal and the external spheres of an organisation's interest is recognition by management that an organisation needs to respond to both primary (those essential to an organisation) and secondary stakeholders (those not essential for an organisation but who can influence it). External drivers include the actions of internationally-mobilised human rights and environmental activists and international and national agreements and regulations concerning environmental protection and social and environmental justice. Internal drivers include management push for *business advantage*, employee awareness, and the drive to accumulate new forms of capital.

The need to accumulate new forms of capital is a recent force for corporate change e.g. companies now trade in reputational capital. Their interaction with consumers and with their workforce is symbolic as well as economic (Beck, 1999; Tsoukas, 1999). Many corporations now market themselves as compliant according to standards set by governments, by negotiated arrangements between government, industry and nongovernment associations or by a range of other non-government associations. The leadership for this change is also coming from managers who recognise the benefits for an organisation in participating in the new coregulatory system i.e. they see *reputational rewards* in voluntarily working towards social and environmental standards which are not enforced by government. Public (and employee) awareness is running ahead of most government and corporate initiatives.

Knowledge management is drawing attention to the value of an organisation's human resources. Motivation, qualifications and commitment, when combined with a significant store of 'corporate memory', are a major asset to the corporation. As prized employees hunt for an organisation with a strong sense of values, there are real rewards in becoming an employer of choice. Organisations need employees who can give high levels of customer service and who are sufficiently motivated and inspired to achieve by the long-range philosophy of an organisation. All industry sectors accept the importance of employee loyalty and commitment (Wilkinson et al, 2001).

Shareholders and investors are also looking to more than financial success in the assessment of performance. Their selection of investments increasingly takes into account reputation and performance on longterm factors of social and ecological sustainability. The growth of the socially responsible investment (SRI) industry reflects this shift. The number of SRI managed funds in Australia has increased substantially. In 1996, there were 10 funds and in 2001, 46 funds. Ethical Investor magazine now lists 89 ethically managed funds. From 2001–2004 SRI funds in Australia have grown by 96%. Strong growth in SRI in the UK has been due to many insurance companies applying SRI criteria across all their equity funds (Sparkes, 2002). Investors are also placing more value on the human capabilities and commitment that the organisation has built. In the new economy the building of knowledge systems, social capital and other strategies designed to increase and sustain human capability are vital to corporate performance.

More and more *employees* have strong expectations for workplace safety and heightened environmental awareness. They are searching for more meaningful work, particularly for work that makes a social and ecological contribution as well as providing an income. In a recent Australian survey, consumers identified the key attributes of the successful company as innovation, care for employees and demonstrable community involvement (Lloyd, 2004).

Concerns about developing *social capital*, can explain why organisations may wish to support local community participation in corporate decision making. Social capital is fundamental to the successful working of the new organisational forms such as the network organisation and communities of practice. Organisations are increasingly dependent on employees who can work cooperatively and contribute to the social capital of the organisation (Adler and Kwon, 2002; Sagawa and Segal, 2000).

18.3 Defining the sustainable culture

We argue that achieving ecological sustainability is dependent upon certain cultural characteristics. This means that the values embedded in the organisation must reflect commitment to the natural environment, employees and communities (Doppelt, 2003). They are those of industrial ecology, an industrial system underpinned by principles of community, interconnectedness and cooperation (Ehrenfeld, 2000). This provides a framework for new levels of resource productivity and generates new strategic directions.

In our view moving towards an environmental ethos requires cultural changes at the level of the human sustainability system, implying changes in relations with both internal and external stakeholders. Human Resource (HR), Industrial Relations (IR), Occupational Health and Safety (OH&S) and community relations aspects of an organisation are therefore implicated in the cultural change.

Our developmental phase model (Table 18.1) indicates how human and ecological sustainability are interrelated developments along the path to sustainability. At each step of the way, new human capabilities along with corporate social responsibility (CSR) characteristics of the organisation enable further progression of ecological sustainability.

18.4 Changing organisational culture through structure

Change, particularly the transformational change required to embed new values in an organisation, will inevitably encounter resistance and perhaps conflict. The first step is to reduce resistance. A change in gover-

Table 18.1   Human and ecological sustainability

Level	Human sustainability	Ecological sustainability
1. Rejection	Employees and subcontractors exploited. Community concerns are rejected outright.	The environment is regarded as a free good to be exploited.
2. Non-Responsiveness	IR a major issue with the emphasis on cost of labour. Financial and technological factors exclude broader social concerns. Training agenda focuses on technical and supervisory training.	Environmental risks, costs, opportunities and imperatives are seen as irrelevant.
3. Compliance	HR functions such as IR, training, Total Quality Management (TQM) are instituted but with little integration between them.	Ecological issues likely to attract strong litigation or strong community action are addressed.
4. Efficiency	Technical and supervisory training augmented with interpersonal skills training. Teamwork encouraged for value-adding as well as cost-saving purposes. External stakeholder relations developed for business benefits.	ISO 14000 integrated with TQM and OH&S systems or other systematic approaches with the aim of achieving eco-efficiencies. Sales of by-products.
5. Strategic Pro-activity	Intellectual and social capital are used to develop strategic advantage through innovation in products/ services. CSR included in training programs, job descriptions and compensation packages.	Proactive environmental strategies are seen as a source of competitive advantage. Organisation signs up for Global Reporting Initiative and/ or other codes.
6. The Sustaining Corporation	Key goals both inside and outside an organisation are the pursuit of equity and human welfare and potential.	An organisation works with society towards ecological renewal.

Source: Modified from Dunphy et al (2003). See also Table 11.3.

nance systems is the best way to reduce resistance and facilitate change for sustainability (Doppelt, 2003). Governance in the sustainable organisation is not focussed on conditions of directive power in the traditional sense, but on developing structures, processes and HR systems which enable control to be exerted through coordination rather than coercion.

Hard organisational change is associated with rational strategies, structural change, radical transformation, leadership and economic goals (Stace and Dunphy, 2001, p. 14). Such a change strategy can forward the sustainability agenda in certain instances. But essential to progress towards an environmental ethos is change to flexible, relatively non-hierarchical structures and decision making. Information gathering processes which help the organisation bridge to other organisations are essential (Clarke, 2004; Doppelt, 2003; Dunphy et al, 2003).

Integration across functional areas will promote environmental progress but is also good business. It points to the need for cross-functional relationships and the building of trust between areas of business that have previously been regarded by managers as only loosely connected (Piasecki, 1995). Active relationships can be fostered between line managers and sustainability experts, so that the issues of human and ecological sustainability are not perceived by employees as irrelevant to the core priorities of management.

18.5 Changing organisational culture through people

To reach the stage of compliance (Level 3, Table 18.1) requires organisations to go beyond changing policies and values to enlist the commitment of all employees and build practical procedures which everyone in the organisation can understand. Structural change provides some of the social organisation needed for the shift to sustainability. But enthusiasm and commitment of the workforce at all levels has to be generated through *soft* change associated with adaptive strategies, cultural changes, continuous improvement, empowerment, and social goals.

Since change for sustainability is dependent upon both *hard* and *soft* change, the human resource function is crucial e.g. an EMS needs to be supported by human resource and community development tools or techniques.

A major barrier to the organisational learning required for the change may be that employees see the natural environment as a technical issue and difficult to understand. Attaining compliance with environmental laws may require new technical solutions or programs being implemented to enable cleaner production techniques and improved worker health and safety. These often include operating within specific, technical considerations such as development consents or specified standards, eliminating hazardous operations, waste disposal, water quality and planning requirements for new developments. Consequently, ongoing HR support may be required if these measures are to be successfully introduced and maintained.

Many managers are recognising that learning capacity is the true capital of an organisation. Dealing effectively with such issues as risk liability is dependent on this capacity. The successful organisation encourages the development of a culture which is sensitive to its general environment (de Geus, 1997). Ongoing improvement in ecological sustainability is assured because the organisation becomes an open system, capable of learning through interaction among and between employees and a wide range of other stakeholders. Implementing environmental efficiencies and responsibility become an integral aspect of the way people conduct business.

Managers are increasingly required to expand the knowledge base of the workforce and institute high performance management practices. Studies demonstrate that these practices (e.g. reduction of status differences, sharing information and extensive training) can build commitment (Pfeffer and Veiga, 1999) i.e. a culture of voluntarism can only be developed through a shared sense of identity. However, this is increasingly difficult to create given that career paths are ill defined and unpredictable and more employment is contractual and part-time rather than full-time. Palmer and Dunford (1997) point out that as a result of these conditions, managers need high-level skills in team-building, empowerment and the development of trust. Because a certain proportion of expert knowledge is tacit, working in self-managed teams requires trust, otherwise the critical knowledge needed to make the new systems work will not be exchanged (Pfeffer and Veiga, 1999).

18.6   Changing attitudes

As surveillance at all levels of the corporation is not possible, changing corporate culture for an environmental ethos requires the internalisation of new standards by people at all levels of the organisation. So attitudinal change is essential to ensure respect for all legislative requirements. Attitudinal changes are required where employees have not been aware of their responsibilities to environmental legislation as members of the organisation or if the collective assumption within the organisation has been that exploitation of community groups, employees or the environment is justified by economic benefits to the corporation. Such change cannot be enforced – it evolves with leadership and over time.

The 'machine' metaphor of the closed and highly regulated system must go if attitudes are to change. This is recognised at a governmental level with the increasing emphasis being placed on self-regulation. With organisations now partly responsible for instigating and monitoring compliance, institutionalising reflection and implementing feedback become priorities. The whole organisation needs to emerge from a legalistic understanding of regulation and begin to support what may seem less tangible ends, such as the capacity of employees to examine their own attitudes towards change.

Limiting staff training to the technical training required for compliance is an add-on approach, not likely to bring about either attitudinal changes or allow the development of a culture capable of adopting self-regulation. Restricting skills development to a few key individuals can also lead to disaster, particularly in some industries where toxic chemicals or dangerous technologies are used by others in the corporation.

While management needs to make explicit reference to new goals, roles and responsibilities to focus attention and energy across the organisation, informal roles can also enable inter-unit collaboration and develop commitment e.g. the champion, the gatekeeper, the idea generator and the sponsor (Adler and Tushman, 1997). These informal roles can become ineffective if formalised, but thrive when those in them are good networkers and strongly committed to the core value of sustainability.

Setting up community consultative committees is one method of fostering the required internal and external cultural changes. Placing employees in face-to-face situations with those who will be most affected by their work activity can have a powerful impact on attitudes. However, although employee empowerment and engagement is a key aspect of these initial changes, both top-down and bottom-up approaches are needed. Neither approach is sufficient on its own (Stace and Dunphy, 2001).

18.7 Indicators to measure culture change

Sustainability indicators measure the progress of organisations toward sustainability. Environmental indicators reflect either the effect of past activities on the organisation (lag indicators) or current activities which may eventually effect the organisation (lead). To measure the culture changes which improve these environmental conditions, requires examining factors such as those that measure adherence to sustainability values across the organisation. Specifically these may be:

- team learning (e.g. group rewards);
- cooperation with a wide range of external stakeholders (e.g. employee days spent volunteering);
- extent to which an organisation is an employer of choice;
- training including environmental or CSR components; and
- extent to which operations breach environmental or human sustainability codes (e.g. OH&S) (Doppelt, 2003; Grayson and Hodges, 2004).

18.8 Roles of environmental change specialists

Environmental specialists play an important role in assisting the direction of the cultural change process in organisations. Although their professional skills may lie in the area of science or engineering their role goes beyond the technical and environmental specialists are increasingly required by legislation and codes of conduct to set their management decisions concerning the environment in the social or political context (Harding, 1998).

Table 18.2   Skills required for change leaders

Change agent characteristic	What it means for leadership
Goal clarity	Knowing what outcome is needed
Role clarity	Knowing when and how to act as a change agent
Relevant knowledge	Understanding the environmental issues in the organisational context – physical, social, political
Relevant competencies	Building continuously the skills and capabilities needed for the change process
Self-esteem	Having the emotional resilience and self-understanding to direct the change

Source: Dunphy et al (2003), pp. 265–266.

One key implication of this shift is towards community consultation and participatory decision making, through such mechanisms as community consultative committees. For instance, a major chemicals company, Orica Pty Ltd (Sydney, Australia) has been required by the NSW Government to consult with two community-based committees in its decision making concerning disposal of highly toxic waste and groundwater contamination associated with the operations of its Botany plant. Arguably, such participation leads to a more open and flexible corporate culture, one that is more able to scope and plan for future risk management issues. As noted, failure of environmental specialists to consult with internal and external stakeholders can be the cause of conflict and resistance to change (Chapter 19).

It is rare for a single change agent to have the full range of skills needed for a complex organisational change program and this is particularly true for change toward environmental sustainability. This suggests the need for a team approach to the implementation of environmental systems. Note that in the internal operation of such teams and in developing the implementation of change with other interest groups inside and outside the organisation, the skills of conflict resolution are vital in resolving tensions without smoothing out the differences we have argued are essential for creativity. The skills of leadership such as communicating a shared vision are vital for gaining commitment to realise opportunities. Recently it has been argued that the 'feminine' style of leadership (Ross-Smith, 2004), based on consultation and communication, can foster the rich diversity associated with the innovative and adaptive culture required to support an environmental ethos.

Environmental specialists cannot remain engaged permanently in the process of consultation. They must also lead towards an outcome. Table 18.2 identifies the skills of effective change leaders.

18.9   Strategies and plans for changing corporate culture

Sustainability and the development of an environmental ethos require the implementation of both environmental and human resource reviews. A stakeholder analysis, which defines roles and responsibilities including those of senior management, is a key aspect of the review. Establishing a structured, systematic response to compliance requirements depends on implementing learning strategies and appointing change agents or specific change units, such as compliance and environmental management committees, human resource and environmental health and safety units to gatekeep information.

To go beyond compliance, these responsibilities need to move from the periphery into the core function of an organisation (Hoffman, 1997) to be integrated into such departments as marketing and product and process design. These reforms will succeed only if employees recognise that the reforms create value for them. Employee ownership of the changes takes us into the area of cultural change within the context of an overall Sustainability Plan. The organisational change strategy developed to achieve the Sustainability Plan must be selected to suit the mission and context of an organisation. Too often environmental managers and human resource managers move in different worlds from strategic business managers. The environmental manager is technically focussed and often unaware of business objectives, except in the broadest sense. Similarly the human resource manager's interest may be in training and other intra-organisational needs. As a result, if sustainability rests in their hands alone, sustainability issues can be marginalised or moved off to agendas unrelated to an organisation's core business. As a counter, environmental policies should be integrated into an overall Sustainability Policy and become an integral part of an organisation's business strategy.

Sharing knowledge across an organisation and developing an integrated policy according to the Triple Bottom Line (social, environmental and economic) requires more coordination between internal elements of an organisation. A significant step along this road is the merging of human and ecological sustainability policies through the explicit recognition by management of the need for a comprehensive sustainability strategy.

18.10   Governance of the change

The relationship between human resource policies and environmental sustainability is reflected in the increasing role played by external organisations in monitoring for sustainability. One reason for this shift is the trend toward flatter organisational structures, voluntary commitment to change, more participative management styles and the implementation of self-managed teams, task forces and project groups. These policies,

Case study: Carlton United Breweries

At the Kent Street site (Sydney, Australia) of Carlton United Breweries (CUB), a major Australian brewer, a well-established learning culture has acted as a base for moves to eco-efficiency. CUB has a Lifelong Learning Program, a team-based approach to organisational learning, work practice and problem solving. Implementation of this Program involved a transformational cultural change (See Table 18.3). In one example with profound implications for environmental management, work area team managers were made responsible for Health, Safety and Environment instead of just focussing on production. As a result, the role of the Environmental Coordinator at CUB is to help teams address their particular issues. A new emphasis on competency related pay scales fostered the move to a systems approach to gain certification through ISO 14000, as well as health and safety and TQM. The raised level of staff engagement and awareness has assisted the gains made in efficiency through automation so that the water required for beer production has halved in 8 years.

Table 18.3 The changing culture at CUB

Old culture	New culture
• Little on-site training, or over award wages,	• Competency based training system linked to salaries,
• Overtime driven,	• Reduction in overtime,
• Every tradesman had an assistant who carried the tools,	• No tradesmen's assistants,
• Overstaffed due to demarcation.	• Major demarcation eliminated.
• Employees lived in local area.	• System of 3 day, 12 hour shifts, with 4 day breaks allows many to relocate further away.
• Layout emphasizes strong gulf between management and workforce (demarcated space).	• Changed layout brings management and workforce around day to day operations (relational space).
• Brewers in offices.	• Brewers out in the workplace.
• Shop steward controls communication.	• Free flowing communication concerning environmental awareness issues and ISO 14000 requirements.
• Most employees only aware of their own functional area.	• Employees familiar with OH&S and environmental issues such as reducing water usage.
• Managers make decisions, • Hierarchical culture with different uniforms for each class of worker.	• Joint decision-making Topic Teams explore and make recommendations for decisions under consultation with Environmental Coordinator
• 19 unions on site.	• Four unions on site.

Case study: Carlton United Breweries – *continued*

Table 18.3   The changing culture at CUB – *continued*

Old culture	New culture
• Conflict and demarcation ridden.	• Consultation e.g. the first enterprise agreement went through over 20 drafts.
• Localised production decisions.	• National production decisions.
• Inward-looking culture.	• Strong learning relationships with stakeholder organisations e.g. Lend Lease CUB's project developer.

Source: Benn et al (2004b).
Note: The views expressed in this paper are those of the authors, not necessarily those of CUB or its officers, either past or present.

---

while increasing employee engagement, can make monitoring according to concrete goals, continuous and relentless improvement and consistently applied performance assessment criteria more difficult. In these conditions external monitoring can enable more systematic assessment and add legitimacy to change programs aimed at ensuring compliance.

A governance system which supports an environmental ethos should prioritise coordination rather than the traditional emphasis on control of delegated power. The key features of the organisation which need to be coordinated are 'access to decision making and information' and the 'distribution of resources'. This requires redesigning the structures, processes and relationships which control access to decision making and information and the distribution of resources. Interpretations of corporate governance need to move from a rules-based perspective to a broader understanding of the structures, processes and relationships which control and coordinate organisations.

18.11   Summary

In this chapter we have described why human sustainability needs to be an integral part of the progress towards ecological sustainability. We have discussed how human sustainability can be achieved through change to organisational structure, people's attitudes and people's behaviour. We have suggested a change strategy linked to business strategy using a phase model with 6 levels to progressively measure change. The importance of environmental specialists as change agents is emphasised.

18.12 Questions

• Summarise the synergies and tensions between human and environmental elements of sustainability, and discuss how these might vary with the three alternative organisational structures discussed in Section 17.4.

• Compare the advantages and risks associated with the two approaches to making cultural change to an environmental ethos – incremental and transformational – and indicate how the organisation roles (discussed in Section 17.6) would be different with the two approaches.

19
Interaction with External Stakeholders

Fiona Court

19.1 Introduction

This chapter discusses the processes involved in interacting with stakeholders and undertaking community consultation. Although the obvious context is new capital projects, many of the principles apply to a business organisation's ongoing operation and strategic planning.

Interaction with stakeholders and the community may be required because:

- it is felt an approvals process external to an organisation would fare better with consultation with possibly affected stakeholders included in project planning;
- there is a legislated requirement specifically for community consultation;
- it is felt this process may help avoid costly challenges later;
- the organisation has a commitment to best practice environmental management that includes the need to consult stakeholders and the community;
- the organisation has a public commitment to transparency and consultation – even if simply a part of its profile management; and
- it is felt the issue is in the public interest, and it is likely the media may also feel that way.

The project stakeholders and communities of interest can: develop biases, readily accept and use inaccurate information, lobby or react with a limited understanding of a proposal, and express anger that their concerns are not being immediately incorporated. If this is the case, conducting a consultation process that has integrity may assist the organisation in relation to their project. There will rarely be consensus but good environmental management is not based on decisions made due to the public's immediate response or the path of least resistance. A good *consultation process* will:

- define your project or proposal in a way that everyone understands;
- create lateral, smart thinking; and

• make sure all the environmental issues are tested thoroughly (which will help you 'sign off' on your project).

In other words a good consultation process that has integrity will help the project stakeholders, the organisation (proponent) and the project itself. A good process, done well, can result in innovation, relevant decision making, positive environmental and social outcomes, and fulfil a corporate commitment to its stakeholders and the community.

19.2 Engaging external stakeholders

A strategy can be developed by considering a series of questions:

• what are the problems or issues the organisation is facing?
• who are the stakeholders in relation to the problems or issues?
• what are the possible parts of the strategy or project that are negotiable?
• what is the objective of the consultation process? (i.e. what is the commitment being made to this community?)

A good step at the beginning of project development is to list each project issue and those people or stakeholders that would have an interest in that issue. There are several levels of stakeholders:

• broad stakeholders – people interested in the strategic considerations of your project. These may be high profile, representative groups or associations;
• local stakeholders – people interested in local level impacts and benefits – at a community level; or
• immediate stakeholders – the project neighbours who are physically or immediately affected by the proposal.

The level of interest from individuals is likely to be highest in the latter category – those closest in time and space (Staib, 1997, p. 40). However, the strategic (and perhaps higher profile) groups in the first category may have more influence at a political level, and as such may need early consideration in project development as well.

Stakeholders can include: Federal, State and local government agencies; Federal, State and local elected representatives; interest groups and associations; local residents and community representatives; project neighbours; media – local and regional; the project team; the approvals organisation. How severe an impact and how concerned these stakeholders are likely to be is a key to determining whether and how much consultation is needed.

19.3 The negotiable parts of a project

The term *project negotiables* refers to those parts of the project that can be changed, adapted, redesigned, tightened or removed etc. in response to environmental and social criteria. Before consultations start the technical team should list and agree on these – and in doing so show they are prepared to be flexible.

Remember that trust starts when a request is made and the project proponent can facilitate a constructive response to meet this concern. Initially, it may not be about the outcome, but about the process. Dialogue must start with establishing a relationship, a connection, a shared understanding. A proponent embarking on a consultation process must be willing to demonstrate a commitment to the consultation process. Remember, a focus on process rather than immediate outcomes…the easiest negotiable to start with is to ask those who indicate they want to be involved a simple question, 'how do you want to be consulted?'

19.4 The objective of the consultation process

The International Association of Public Participation (IAP2, 2000b) developed a spectrum of participation that helps to define what promises the organisations' leaders are prepared to make. Is the organisation going to: inform, consult, involve, collaborate with, or empower the participants (Figure 19.1)? Each of these levels of activity is worthwhile, so long as there is honesty about the objective that is being adopted. In this way the approach differs from 'Arnstein's Ladder' discussed in Section 9.5. Publicly stating a promise to a process resolves the problem of unreal expectations being created about the level of consultation and decision making.

19.5 The barriers to community consultation

Equity in consultation is an important principle for access to information and for opportunities to be involved. The main barriers to involvement in a consultation process that need to be considered during the design of a consultation program are listed in Table 19.1.

19.6 Steps to engage with external stakeholders

The key steps to commencing your dialogue with stakeholders are to:

- prepare a community and stakeholder *involvement plan*;
- prepare some plain *English tools* to help with information dissemination and aid two-way discussion;

IAP2 Public Participation Spectrum

Developed by the International Association for Public Participation

INCREASING LEVEL OF PUBLIC IMPACT ➤

INFORM	CONSULT	INVOLVE	COLLABORATE	EMPOWER
Public Participation Goal:	**Public Participation Goal:**	**Public Participation Goal:**	**Public Participation Goal:**	**Public Participation Goal:**
To provide the public with balanced and objective information to assist them in understanding the problems, alternatives and/or solutions.	To obtain public feedback on analysis, alternatives and/or decisions.	To work directly with the public throughout the process to ensure that public issues and concerns are consistently understood and considered.	To partner with the public in each aspect of the decision including the development of alternatives and the identification of the preferred solution.	To place final decision-making in the hands of the public.
Promise to the Public:	**Promise to the Public:**	**Promise to the Public:**	**Promise to the Public:**	**Promise to the Public:**
We will keep You informed.	We will keep you informed, listen to and acknowledge concerns and provide feedback on how public input influenced the decision.	We will work with you to ensure that your concerns and issues are directly reflected in the alternatives developed and provide feedback on how public input influenced the decision.	We will look to you for direct advice and innovation in formulating solutions and incorporate your advice and recommendations into the decisions to the maximum extent possible.	We will implement what you decide.
Example Tools:	**Example Tools:**	**Example Tools:**	**Example Tools:**	**Example Tools:**
● Fact sheets ● Web Sites ● Open houses	● Public comment ● Focus groups ● Surveys ● Public meetings	● Workshops ● Deliberate polling	● Citizen Advisory Committees ● Consensus-building ● Participatory decision-making	● Citizen juries ● Ballots ● Delegated decisions

© 2000 International Association for Public Participation

Figure 19.1   The IAP2 Public Participation Spectrum
Source: Reproduced with permission of International Association of Public Participation (IAP2, 2000b).

- plan how your *technical team* and the decision-makers will be involved in consultation processes, to ensure they understand community issues first hand; and
- understand when and how to *finish* the program.

Table 19.1   Possible barriers to participation

Who	Could be restricted by
Women	Home care or childcare responsibilities, sexism
Children and young people	Peer group pressure, feelings of inadequacy, use of jargon or English that is above the comprehension level of the participants
Older peoples	Lack of access, e.g. transport or cultural differences
Indigenous peoples	Lack of familiarity (may feel uneasy in groups in which they are not familiar), racism
Single parents	Time, resources
People from non-English speaking backgrounds	Cultural differences, may feel uneasy in groups in which they are not familiar, may not be fluent in English, use of jargon
People with a disability	Access, may feel uneasy in groups in which they are not familiar, resources
People living in remote or rural areas	Lack of immediate access, conflicting priorities, e.g. changing work schedules depending on external factors such as rain, difficult to make aware of the project given reduced media outlets
Shift workers	Activities programmed during those times they are at work

A community and stakeholder involvement plan

Using the information developed from earlier sections in this Chapter on: what is the problem; who are the stakeholders; what are the possible parts of the strategy or project that are negotiable; and what is the objective of the consultation process, a plan can be prepared. This plan should address the four questions raised in Section 19.2. The plan can be complex (for a very large cost project with multiple stakeholders), or quite simple (for a straight forward project). It should identify: project context; consultation objectives; stakeholder analysis; the approach to activities and why certain tools are chosen; a schedule of planned activities against timeframes; and roles and responsibilities.

Preparing a plan should be a team effort with the proponent team and internal decision-makers examining together the likely risks and barriers to the successful development of a project. This requires canvassing all the likely stakeholder issues, and ways to manage those processes such as: providing specific up front information on an issue; getting external or independent peer review; forming a consultative committee to monitor research or evaluation processes; being prepared to negotiate on particular aspects of the design to meet the needs of this issue; or hiring a specialist.

Plain English consultation tools

Tools must use plain English – to both widen your readership and also to assist your own thinking e.g. what is the project and what the reasoning is behind what you have adopted? There are many ways to engage with stakeholders, from workshops to committees, and notice boards to radio announcements. Figure 19.2 shows a range of possible tools. Tools are needed to:

- create awareness of the proposal;
- create ways to have two-way discussion about a project;
- ensure that the ideas, options, issues and solutions are incorporated into the work of the technical teams; and
- provide feedback to the stakeholders or wider community as to what issues or ideas could be incorporated, and what could not, and why not.

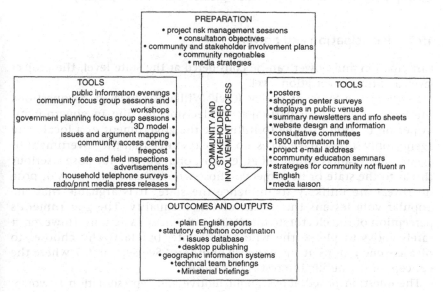

Figure 19.2 Tools to use in the consultation process

The technical team

Technical specialists are often reluctant to incorporate community issues into their designs. This reference to 'technical' includes studies that are scientific (relying on quantitative data) and other disciplines based on values and qualitative data: social impact, economic, urban design. In all cases these specialists will be protective of the chosen methodology – understandably as the science used in an evaluation relates to the integrity of the scientist, and community involvement may or may not improve the method or level of assessment.

In narrating the development and subsequent compromise of the Alborg Town Plan, Flyvbjerg (1998) describes how those with political power and social standing in the public debate led to the rationalisation of scientific method. Flyvbjerg's premise is that modernity relies on *rationality* as the main means for making democracy work, rationality in decision making in accordance with scientific principle. His book questions the quality of the outcomes resulting from the trend towards increasing participatory processes.

A level of flexibility is sometimes required – usually going beyond what a scientist may deem as necessary, e.g. obtaining independent peer review. This is a part of the consultation process that seeks to build trust. The technical specialists need to be incorporated into the consultation process. This will reduce an inherent reluctance for flexibility. It can assist their understanding of key concerns, the need for method change, the need to reconsider assumptions or the need to provide a greater level of transparency.

19.7 Participation and power

Participation and power can be considered at the state level, the project level, and the conversation level.

Firstly at the State or policy level. Williams (2004, p. 5) questions the level of equity in participatory decision making in government. 'A person or group is able to influence the government or a local state agency only if they have access to some resource that the government or agency wants and cannot get at less cost or at all, or they pose a serious threat to the state or its projects. Those who hold no resources or pose no threat are unlikely to influence the state. It is arguable that the popular vote lessens the extent of this inequality. The government's perception of the electorate inevitably influences its actions. However, it rarely looks to please the whole electorate, but tactically chooses to please some groups at the expense of others.' See Section 9.7 where the concept of the public interest is introduced.

The question posed above about motive and representation is worthwhile considering. It may be less of an issue for private organisations in their dialogue with communities and for individual questions of environmental policy and management. However, it is an important question in the establishment of environmental regulations and major environmental policy.

Secondly, the issue of a project using the community involvement process to provide a symbolic tick for the approvals process. Painter (1992, pp. 32–35) provides an excellent description of the participation process and the issues of power sharing and the potential loss of community negotiating power as a result of the consultation process. ' For the participants in this sort of process, "buying in" is always accompanied by a process of "buying off". However, this is a two-way process with opportunities and risks on both sides. Once an authority structure has committed itself to consultation or "sponsored" participation, it has set in train a process that it may not be able to manage. The outcomes of information sharing and negotiation with outside groups are unpredictable. On the other hand, the outside group must consider the potential traps, such as the possibility the authority structure will use the consultative exercise as a subsequent legitimisation for its own decision.'

Some groups will choose to remain 'on the sidelines' where their lobbying and argument may be more powerful – and have a greater chance of producing change.

Thirdly, power at the conversation level. Upon learning of a proposal, communities will have very little information (unless providing information is a specific strategy of the proponent). To fill the gap in of this feeling of powerlessness, groups without information will seek it – readily available through the internet – be it good science or bad. Providing all relevant information *upfront* and being transparent with information as it is collated can partly resolve this issue. Waiting until a project report has been finalised and issued can lead to frustration (time delay is often a tactic to reduce the level of outrage and response) and community opinion leaders may take a stance without waiting for that information. When the 'good science' is released, these community opinion leaders may not be able to move away from their publicly stated positions.

19.8 Making a commitment to a local community

The process of making a commitment to a consultation process that has integrity, or is seen to be credible by the community is outlined in Figure 19.3.

19.9 Principles for effective consultation

Documentation on best practice consultation includes manuals such as the NSW RTA (1998) and Carson and Gelber (2001). These manuals include evaluation criteria or principles that can be used to measure or evaluate the consultation process. The consultation process can be assessed by determining whether it was:

- efficient and accountable;
- equitable and inclusive (did it overcome barriers to involvement);
- effective – a process that allowed real participation (realistic access to information, did it allow enough time, were the results reflected in decision making);
- flexible – responding to the needs of those involved;
- fair – conducted with integrity and subject to evaluation;
- a process that achieved representative involvement from a range of interests;
- cost-effective – reflecting the size of the program against the level of interest in the issue; and
- interactive and deliberative – was dialogue and debate achieved and was the process well-facilitated.

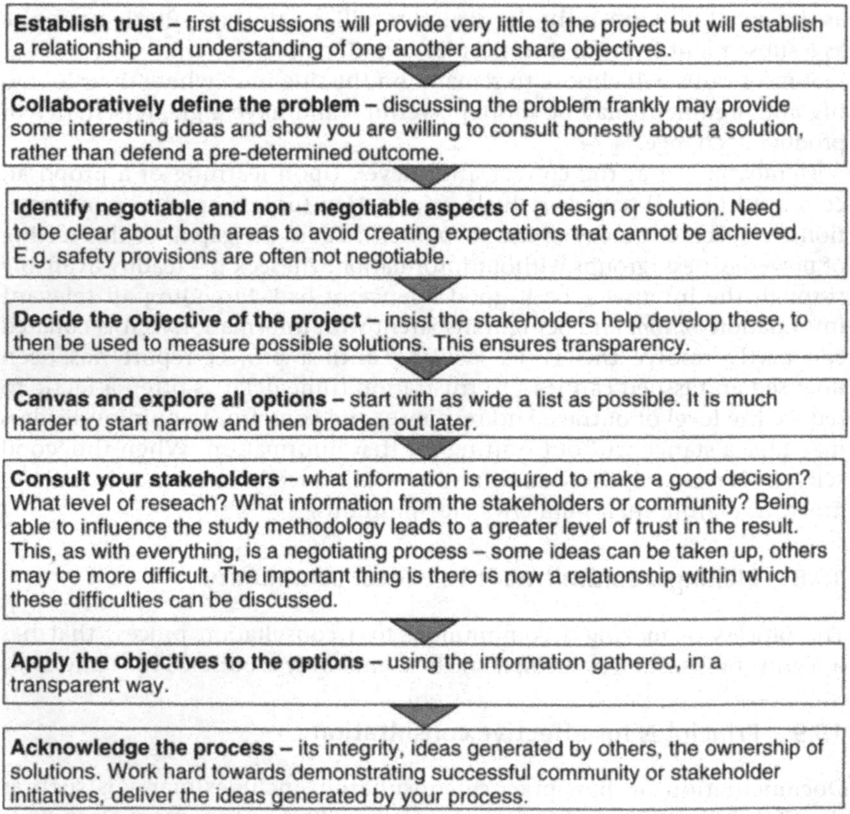

Establish trust – first discussions will provide very little to the project but will establish a relationship and understanding of one another and share objectives.

Collaboratively define the problem – discussing the problem frankly may provide some interesting ideas and show you are willing to consult honestly about a solution, rather than defend a pre-determined outcome.

Identify negotiable and non – negotiable aspects of a design or solution. Need to be clear about both areas to avoid creating expectations that cannot be achieved. E.g. safety provisions are often not negotiable.

Decide the objective of the project – insist the stakeholders help develop these, to then be used to measure possible solutions. This ensures transparency.

Canvas and explore all options – start with as wide a list as possible. It is much harder to start narrow and then broaden out later.

Consult your stakeholders – what information is required to make a good decision? What level of reseach? What information from the stakeholders or community? Being able to influence the study methodology leads to a greater level of trust in the result. This, as with everything, is a negotiating process – some ideas can be taken up, others may be more difficult. The important thing is there is now a relationship within which these difficulties can be discussed.

Apply the objectives to the options – using the information gathered, in a transparent way.

Acknowledge the process – its integrity, ideas generated by others, the ownership of solutions. Work hard towards demonstrating successful community or stakeholder initiatives, deliver the ideas generated by your process.

Figure 19.3 The commitment to community consultation, involvement and collaboration

Although decision making can strive for consensus, complete agreement almost never occurs – there is always a range of competing values and interests. However, if the process was a transparent one, involved an energetic and responsive dialogue, and allowed participants to see *how* a decision was reached, then many will feel satisfied even if the outcome was not as they would have completely desired.

19.10 Conclusions

Some final communication and consultation tips are:

- community consultation should not (no matter how tempting) be used as a way to capture good news headlines in the media; the process of dialogue must be a mutually agreed one.
- good strategic reasoning is needed before you start consultation (it is not a quick fix) and without good science to begin with, the project will unravel under public scrutiny).
- clearly recognise the legitimacy of community consultation as part of the wider or larger planning process.
- try to understand the wider context within which the strategy or project is being proposed – what other issues are your stakeholders facing and how are they likely to influence the response to the project being examined?

Mackay (1994) outlines further communication tips for talking with people. In relation to involving stakeholders in environmental outcomes they are invaluable. They help explain how key messages are interpreted. *Firstly*, it's not what our message does to the listener, but what the listener does with our message, that determines our success as communicators. *Secondly*, listeners generally interpret messages in ways that make them feel comfortable and secure and conversely, messages that make people feel uncomfortable will be often misinterpreted. *Thirdly*, when people's attitudes are attacked head on, they are likely to defend these attitudes, and in the process to reinforce them' ... much as Thomas Paine stated in 1792 (Paine, 2004).

Part IV

Environmental Management Techniques and Tools

20
Environmental Decision Making
Robert Staib

20.1 Introduction

In this chapter we discuss the concept of hierarchical or sequenced decisions, the decision making process, the framework in which decisions are made, how to structure decisions for ease of analysis and finally make reference to some of the tools that can assist the decision maker. All the while there is the need to consider how the environmental aspects can be best introduced and used in organisational decision analysis and decision making.

Decision making is a thread that winds its way throughout this book and also through environmental management. Many texts on environmental decision making focus on the political and governmental approaches to decision making (See Part I of this book), whereas in this chapter we discuss decision making from an organisational point of view.

Organisations are continually making decisions from the strategic to the daily. These may be concerned with: the products they sell or wish to sell; the materials, supplies and equipment to be used; the technology to apply; the location of the next manufacturing or distribution facility; how to package, transport and price their products; and how to comply with new laws and regulations (Chechile and Carlisle, 1991, pp. 238–244)

Organisational decisions can be made progressively over time, using different processes and using many different criteria and can be by individuals or groups (Fulop and Linstead, 1999, Chapter 8). The decision maker or makers can be supported by a decision analysis process linked to different types of decision aids or tools (Clemen and Reilly, 2001, p. 3). The outcome of the decision making process is the decision to be implemented.

With increasing pressure from governmental environmental regulation, customer and community concern, and a growing internal environmental awareness, organisations need to introduce environmental factors into their decision making processes.

20.2 Hierarchal or sequential decisions

For many years, project management has used the concept of the hierarchical or sequential decisions e.g. decisions made early in a project have a larger impact on cost than those made later in a project (Staib, 1999). A similar concept can exist with decisions and environmental impacts.

The Sydney 2000 Olympics provides an instructive example. The decision to hold (or need for) an Olympic Games potentially creates and embodies large environmental impacts through the use of resources and energy and the generation of pollution. The decision by Australia to offer a 'green' 2000 Olympics changed the normal scope of an Olympics and consequently reduced the potential environmental impact. Later design decisions by the Sydney Olympic Coordination Authority mitigated this environmental impact e.g. planning and building facilities for buses and trains greatly reduced the reliance on private cars, bringing with it reductions in the potential impact of energy use, resource use and pollution. A third decision in the hierarchy, the decision to use recycled timber on the station handrails created an environmental benefit though its magnitude was small in comparison (Olympic Co-ordination Authority, 1997; Staib, 1999).

The hierarchy of decisions for the Olympics project progressively reduced the potential environmental impact but did not eliminate it. This concept is shown diagrammatically in Figure 20.1 and illustrates that early decisions in a project or program are likely to produce greater environmental mitigation than later ones. (The axis in the figure is not to scale – though it should be possible to develop a quantitative graph.) These temporal aspects can also be considered as a sequential decision making process (Clemen

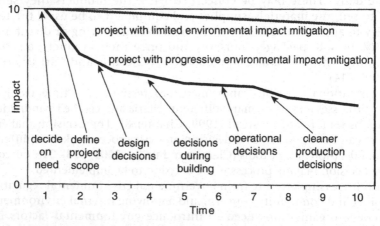

Figure 20.1 Decision making hierarchy (or sequence) for a project or program

and Reilly, 2001, p. 26) though the hierarchical description acknowledges the organisational hierarchy involvement in the decision making process.

20.3 Environmental decisions: societal versus organisational

A similar environmental impact hierarchy exists within and across societies' institutions. This concept (illustrated in Table 20.1) starts with the political and leads to the organisational level.

This chapter is mainly concerned with the organisation (fourth category) though the political, governmental and consumer decisions represent the milieux in which the organisation makes its own decisions. It is important for organisations to recognise the external and internal hierarchies and sequences. They suggest that the earlier in a process the environmental aspects (impact mitigation or introduction of positive environmental benefits) are considered, the greater should be the environmental benefits. This has implications for decisions made by Boards and Chief Executive Officers.

Table 20.1   Hierarchy of decisions that affect environmental outcomes

Decision arena	Ch *	Types of decisions	Environmental consequences for an organisation
Political	5	Passing of legislation. Directions to government.	Mandatory environmental measures Environmental policy and regulations
Governmental Administration	6	Environmental projects. Constraints on new projects. Operational implementation.	Environmental aspects to incorporate into tenders and contracts. Environmental limits on scope for private projects. Operational management of resources, pollution, waste.
Consumers and customers	9	Purchasing. Reaction to company operations.	Reflects on company sales Reflects on company products and image.
Private organisations	11	Strategic. Projects. Operations.	Affects company environmental directions and achievements. Affects later environmental operations of a facility. Affects consumer purchasing & reactions.

* Chapter of this book.

Environmental decision making in the external arena i.e. political and governmental is broadly discussed in the other chapters of this book as noted in Table 20.1 and is addressed in more detail in specialist books (Chechile and Carlisle, 1991; Harding, 1998).

20.4   Decision making concepts and processes

Rational approach

Many writers describe the rational approach to decision making as one that proceeds logically from the problem or issue through analysis to a decision (Fulop and Linstead, 1999, p. 299; Clemen and Reilly, 2001, p. 6; Lee et al, 1999, p. 4). This is the point where we start our discussion on the decision making process. Although it is a simplified process (Lee et al, 1999, p. 4), it forms a good basis for structuring decision making and analysis. Figure 20.2 illustrates in its centre stream of activities, those that lead up to the point of making the decision. The outcome is a recommendation for the decision maker (an individual or a group e.g. Board or committee) to make the decision. It could also be an organisational hierarchy of people as the recommendation heads upwards in the organisational structure until the person with the appropriate level of delegation signs off and accepts responsibility for it (Chechile and Carlisle, 1991, p. 249).

The top two boxes on either side of the centre stream represent external and internal factors that need to be considered in decision making. Their influence can be strong enough to make rational decision making in an organisation a difficult if not impossible task (Lee et al, 1999, p. 4). If environmental aspects are to be given appropriate weight they need to be considered in all activities as shown in Figure 20.2.

Lee et al (1999) says that the rational process requires that: 'the problem and goals are clearly defined; all alternatives and outcomes are known; preferences are clear, constant and stable; there are no constraints; and final choice will maximise results for the individual and the organisation.' He says that some of the reasons why decision making may not be rational are: 'imprecise definition of the problem; lack of alternatives; no clear criteria; time and cost constraints; an inability to calculate the optimal choice; imperfections of the decision makers' perceptions; incompatibility between attitudes.'

Administrative and incremental approaches

Some authors postulate less structured types of decision making models that they believe more likely reflect reality. These take into account some of these internal influences e.g. *administrative decision making* with less than perfect information and with some form of consensus producing *best in the circumstances* (Fulop and Linstead, 1999, p. 308) or various forms of incremental decision making that are driven by the organisational players and

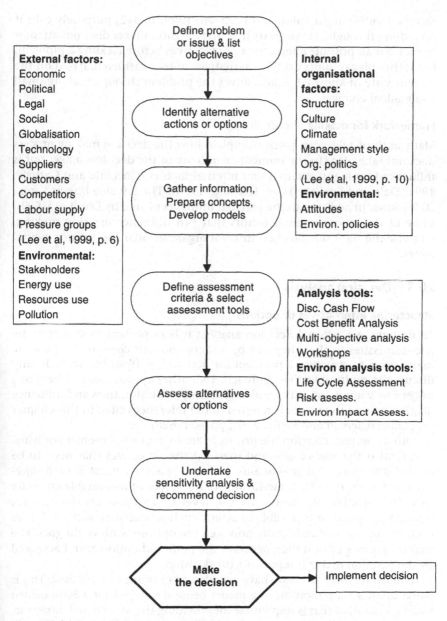

Figure 20.2 Rational decision making processes
Source: Based on Fulop and Linstead (1999, p. 307); Clemen and Reilly (2001, p. 6); Lee et al (1999, p. 4).

factors. Lindblom (in Fulop and Linstead, 1999, p. 312) purposely calls it 'muddling through'. He suggests that 'the decision maker does not attempt to root out all possible alternatives or objectives before tackling a problem, but rather places limits on the alternatives to be considered, based on the current state of knowledge, and solves the problem through making small and gradual changes.'

Framework for making decisions

Many authors take great pains to explain how the decision making process does not take place in the immediate milieux of the decision maker but is influenced by many external and internal factors (Chechile and Carlisle, 1991; Dale and English, 1998; Harding, 1998). The left side box in Figure 20.2 shows, in addition to the external influences listed by Lee et al (1999), some of the environmental factors that can influence decision making. Likewise the right side box lists internal organisational and environmental issues.

20.5   Decision analysis

Structuring alternatives or options

To undertake a rational decision analysis it is important to structure the decision issues so that they can be understood and compared. There are many ways to structure a problem for decision analysis but we will only discuss the multi-objective approach. Two other approaches can help one to get one's mind around complex problems – decision trees and influence diagrams – and the reader is referred to the references cited in this chapter for details (Clemen and Reilly, 2001, pp. 621–628).

With a complex decision the first steps are to define the problem or issue, to establish the *goal* or aim and then list the objectives that need to be achieved to reach that goal or aim. To measure achievement of each *objective* it is necessary to then ascribe one or a number of measurable *attributes* to each objective. By measuring the level of all these attributes for a number of options it is possible to determine how well each *option* achieves each of the *objective* and finally how well the options achieve the *goal*. The highest scoring option then becomes the preferred option and if accepted by the decision maker it represents the decision.

To illustrate this process we have included an example in Table 20.2. This is based upon a simplification of a model being developed for a State owned water corporation that is responsible for providing the water based infrastructure to new urban development areas in Sydney (CSIRO, 2002). In the example the goal is to: 'Achieve sustainable water use in a new urban development'. This is an absolute goal that may be difficult to achieve in the short term. Probably a more realistic medium term *goal* might be to: 'Select the option that best meets the project objectives: financial, social and environmental'. This

Table 20.2   Hierarchy of decisions – example

Goal	W %		Objectives	w %	Attributes
		A	Minimise resource usage	a	Fresh water
				b	Fuel
				c	Material amounts
	E			d	Appropriate materials
	N				
	V	B	Minimise waste	a	To sewer system
	I			b	Sewage overflows
	R			c	From sewer
	O			d	Max. reuse nutrients
	N			e	Max. reuse carbon
	M			f	Other waste
	E	C	Maintain ecological	a	Endangered species
ACHIEVE	**N**		functions	b	Waterways
SUSTAINABLE	**T**			c	Riparian vegetation
	A			d	Land vegetation
WATER USE IN	**L**				
A NEW URBAN		D	Foster awareness	a	Community
DEVELOPMENT	**S**			b	Residents
	O	E	Contribute to amenity	a	Visual
	C			b	Odour
	I			c	Involvement
	A			d	Health & safety
	L			e	Access to land
	U	F	Satisfy utility	a	Reliable service
	T			b	Water quality
	I			c	Water pressure, flow
	L			d	Equal access to service
	I				
	T				
	Y				
	C	G	Minimise cost	a	Life cycle
	O			b	Pricing
	S			c	Economic
	T				

Source: Based upon CSIRO (2002). W = weighting of objectives where ΣW= 100%; w = weighting of attributes and for each objective, Σw = 100%.

aside, it suffices for our demonstration. It takes into account the three main aspects of the anthropogenic urban water cycle: potable water supply; sewage collection and treatment; and management of stormwater quality and quantity. It identifies 7 *objectives* that need to be satisfied if one is going to achieve the *goal*. The objectives are supported by 28 *attributes* spread unevenly over the objectives. Traditionally utility and cost would have been the main objectives but with a goal of sustainability there is the need to incorporate the social and environmental aspects.

There are often many ways to achieve an objective e.g. for the objective (A. minimise resource usage) and the attribute (a. fresh water). Fresh water is traditionally distributed to a city from a large dam external to the city through a system of pipes, reservoirs, pumps and treatment plants. Other ways of supplying fresh water include collection of rain water in tanks on individual properties or groups of properties or for individual high rise buildings, use of recycled treated sewage effluent or use of recycled treated stormwater or a combination of the above.

One needs to consider all the ways of satisfying each *objective/attribute* combination i.e. 7 objectives and 28 attributes. This would end up being an enormous number of mathematically possible options. In practice it is necessary to apply judgement to the selection of options based on a selection of the available and practicable ways of achieving each objective. In the above example ten options were considered for analysis.

Weighting of attributes and objectives, measurement of attributes

After the decision problem has been logically structured, next comes the difficult issues of deciding how much each *attributes* contributes to meeting of its *objective*, deciding how much each objective contributes to meeting of the *goal* and combining the different measurement scales for each of the attributes (Clemen and Reilly, 2001, p. 623; Foreman and Selly, circa 2004).

The simplest approach is to give each objective and each attribute equal weighting i.e. $WA=WB=WC=WD=WE=WF=WG=1/7$ and for objective B *minimise waste* $wa=wb=wc=wd=we=wf=1/6$.

For the hierarchical multi-objective approach to be applied the *attributes* need to be not only measured but need to be added. But as can be seen from Table 20.2, the attributes are going to be measured in different units e.g. fresh water in kilolitres per person per day, health and safety in days of illness or injury, life cycle costs in money values. In order to be able to develop a total score for each option, one must be able to normalise each attribute (e.g. put on a scale of 1 to 10) and multiply it by a weighting (w) and add each product to achieve a total score for each objective.

After the objective scores are arrived at each objective is multiplied by its weighting (W) and added to achieve a total score for the goal against each option. After one has done this for each option, a ranking of options can be determined.

While this method of decision analysis is a much-used tool it is has a number of traps for the beginner and experienced alike and a considerable amount of judgement (and decision making) is required to arrive at a workable model. The key areas are deciding on:

i.   the number and make-up of options to assess;
ii. the method of converting the attribute measurements to a common scale often called normalising;
iii. the weighting (w) to be used for each attribute; and
iv. the weighting (W) to be used for each objective.

The literature includes many methods for developing these aspects, e.g. Clemen and Reilly (2001); Foreman and Selly (circa 2004). The use of specialists in a workshop situation is often used to arrive at a consensus based on each individual's knowledge and experience. The decision analysis is made more robust by undertaking a sensitivity analysis which assesses the effect on the ranking by varying each of the key parameters e.g. i to iv above. This has been a simplified introduction to multi-objective decision analysis and the reader is referred to the references for more detailed explanations of the method, worked examples and discussions of the shortcomings.

For those who like diagrams, Figure 20.3 shows how a decision problem can be structured to include each option. Table 20.2 excluded the options for simplicity. For clarity of presentation Figure 20.3 uses only three levels: the highest level – the primary aim or *goal*; the second level – the *objectives* (in this case 3 objectives); and the third level – the *attributes* (in this case 6).

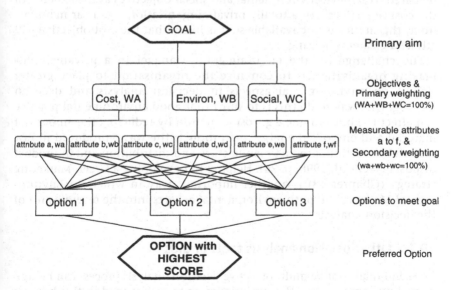

Figure 20.3   Multi-objective decision analysis
Source: Based on Clemen and Reilly (2001, p. 598); Foreman and Selly (circa 2004).

Three options are shown. There is no limit to the number of options, the number of levels, the number of objectives or the number of attributes, but the lowest level of attribute needs to be measurable.

20.6 Environmental influence

The objective types shown in Figure 20.3 are the elements of sustainable development. The financial aspect of decision analysis is very important for the longevity of a private profit-making organisation so the financial objectives tend to dominate. Therefore to introduce the social and environmental aspects needs not only determination but also a certain amount of skill.

The example in Table 20.2 shows the objectives split into 4 categories: environmental, social, utility and cost and is an attempt to measure whether individual options are likely to 'achieve sustainable water use in a new urban development'. Often the objective of the decision analysis process is to find the best option acknowledging that it is not always possible other than for a simple decision to achieve all objectives. There is inevitably a trade off between objectives. While the example in Table 20.2 is seeking the best option it is also seeks an absolute goal of implementing a water system that achieves sustainability.

The example (Table 20.2) is for a State owned water corporation and the Government would have the ability to influence objectives and possibly be prepared to achieve environmental and social objectives at the expense of the cost objectives. For a totally private organisation e.g. a car manufacturer, this ability is not available so it is likely that the cost objective will attract a greater weighting.

The challenge for the Environmental Manager in a private profit-making organisation is to convince the organisation to place greater weight on environmental aspects in decision analysis and decision making. One way to do this is to show that good environmental practice can effect the bottom line e.g. cost reduction by reducing: the amount of material and embodied energy in a product; the waste produced in raw material extraction, product manufacture and distribution and in product disposal (Welford, 1998, p. 26). Helping to developing a Green Marketing strategy (Chapter 13) is another important way in which an Environmental Manager can push environmental aspects into the revenue area of the decision analysis.

20.7 Other decision analysis tools or aids

The semi-quantitative multi-objective decision analysis process can be supported by some of the other techniques or tools outlined in this book as shown in Table 20.3.

Table 20.3 Other techniques in which to include environmental aspects

Tool or technique	Ch.	Use in decision analysis	Environmental aspects
Cost benefit analysis (CBA)	6	By government departments deciding between alternative development strategies or projects	Consider all environmental benefits e.g. less air pollution and all environmental impacts e.g. greater use of non-renewable resources
Environmental economics	8	By government departments to value environmental benefits and impacts	Value of environmental aspects can be added to CBA above.
Discounted cash flow (DCF)	6	Organisational financial analysis of strategies, projects, products	Cost environmental savings in material, energy, waste reduction.
Life cycle assessment	24	By organisations deciding on best product or process with minimum environmental impact	As stand-alone tool or as part of a multi-objective decision analysis.

20.8 Conclusion

In this chapter we have discussed how decision making can be supported by the use of a formal decision analysis process using various tools to assist the decision maker. In making decisions organisations have to consider many issues both internal and external. From an environmental perspective we have discussed the need to always incorporate the environment into the decision analysis process and into the making of organisational decisions and to introduce it at the earliest possible time for maximum benefit. It is not easy since the financial aspects of organisations can be dominant but many techniques are being developed that enable organisations to consider environmental aspects. Some of these can be used to quantify the environmental aspects in monetary terms and be shown to directly affect the bottom line.

Some of the external aspects e.g. a degrading environment, legislation, customer preferences, stakeholder concerns have both cost and revenue implications for an organisation. This is driving the need for organisations to consider all environmental issues in its decision analysis and decision making processes. Failure to do so may make it hard for an organisation to survive and be organisationally sustainable in the medium to long term (Dunphy et al, 2003, pp. 29–84).

Case study: which vehicle is the greenest?

In recent years carmakers in many countries have started to produce and sell cars and other vehicles with a significantly reduced impact on the environment in response, inter alia, to the projected changes to more stringent emission limitations on noxious polluting substances and greenhouse gas emissions (Hawken et al, 2000, p. 22; Orecchini and Sabatini, 2003; Van Mierlo et al, 2003).

A decision faced by organisations (or individuals) is how to select the most environmentally friendly car. Other considerations in the decision may include the life cycle cost (purchase price, on costs including insurance, maintenance cost, resale value), safety and performance. Marketing and research departments may wish to determine if their car compares well with others on (or about to be on) the market.

Van Mierlo et al (2003) describes a detailed life cycle assessment approach to green vehicle selection for people living and driving in Brussels that identifies possible emissions from vehicles including their fuel production and component manufacture cycles (25 chemicals) and their impact on people's health, buildings and ecosystems. The methodology compares 24 vehicles with a variety of fuels: compressed natural gas, liquefied petroleum gas, petrol, diesel and electricity (with electricity from renewable and non-renewable sources); and a variety of drive trains including hybrid vehicles.

The *goal* is to select the greenest car, the *objectives* cover impact on the natural environment and on human health and they include a weighting so the scored attributes can be added. The different types of environmental damage (the *attributes*) are scored for each vehicle and are then normalised by assessing how far the scores are from a reference vehicle. This enables the different types of damage to be added after appropriate weighting. The normalised damage values are weighted by a panel. The addition of these enables one to arrive at an Ecoscore for each vehicle, which can then be ranked. This score could then be entered into a separate decision model with the other objectives (price, safety and performance), weighted and added to determine the optimum vehicle for an organisation's fleet.

20.9   Questions

- In an organisation in which you have worked, who makes the decisions that have the greatest effect on the environment positive or negative?
- Do decisions made at the coalface have a real impact on the use of environmental resources e.g. energy and materials?
- Is a formal decision analysis process undertaken for major investment decisions by the organisation or are the decisions made by the executives based on experience and gut-reaction?
- How often are all environmental aspects considered in a major organisational investment decision?
- Using Figure 20.3 and Table 20.2 as a model, establish a goal, objectives, attributes, weightings (W and w) and normalising process for attributes for the introduction of a new car model that is aimed at the green market.

21
Environmental Risk Management

Rory Sullivan

21.1 Introduction

This chapter provides an overview of the manner in which organisations manage and control environmental risks as a part of their overall management efforts.

Environmental catastrophes such the 1976 Seveso dioxin release in Italy, Union Carbide's accident at Bhopal in India in 1984, the grounding and sinking of the Exxon Valdez oil tanker in 1989 and the Baie Mara cyanide spill in Romania in 2000 have pushed the subject of environmental risk into the public arena. While environmental disasters have received significant media attention, organisational risk management deals mainly with the larger number of lower consequence events.

21.2 Definitions and key concepts in risk

Hazard is defined as the source of potential harm (Standards Australia, 2004, p. 3). From an environmental perspective, hazards can be considered as physical situations with the potential for human injury, damage to property, harm to the environment or some combination of these, as a consequence of events such as fires, explosions, accidental spills or failures of pollution control equipment (Sullivan, 1998). *Risk* is defined as the probability of something happening that will have an impact on objectives (Standards Australia, 2004, p. 4). Risk is measured in terms of *frequency* (or likelihood of occurrence) and *consequences*.

In this context, risk management relates to the systems and processes that are directed towards minimising adverse effects or maximising the opportunities associated with risk. That is the risk management process should provide organisations with a rigorous and robust basis for decision making and planning through the better identification of opportunities and threats. This, in turn, should allow organisations to make more effective use of their resources, improve their incident management systems, enhance stakeholder trust and improve compliance with relevant legislation (Standards Australia, 2004, pp. 1–2).

The main elements of any risk management process are:

- establish the context;
- identify risks;
- analyse risks;
- assess and prioritise risks;
- treat risks;
- monitor and review; and
- documentation.

In the following sections, the manner in which each of these elements apply to the management of environmental risks is discussed.

21.3 Context establishment

The starting point for any risk management process is to understand the context within which an organisation is operating. In general, this requires that the organisation defines its short-term and long-term policies, goals and objectives, taking into account the competitive, political and legal aspects of its operations. For environmental risks, attention should be paid to stakeholder views (e.g. local communities, environmental groups), relevant legal requirements (e.g. due diligence, potential changes in environmental legislation) and the positive and negative financial impacts associated with environmental issues.

Defining the context should also include consideration of the scope of the organisation's environmental risk management efforts. Reflecting changes in the manner in which companies are being held responsible for their actions and for the actions of others, organisations need to consider: past, present and future activities; upstream, internal and downstream impacts; and normal, abnormal and accident conditions. This reflects changes in the manner in which companies are being held responsible for their past actions and for the actions of others. For example, in relation to past activities, it is no longer acceptable for an industrial organisation to claim that, because there was no prohibition on past activity that has pol-luted soil or groundwater, the organisation should not be required to clean up. Similarly, there has been litigation where manufacturers have been held liable for the manner in which their products have been used.

Based on the context definition, the goals and objectives of the environ-mental risk management process should be defined. The goals and objec-tives adopted will vary depending on the nature of the organisation's operations, but could include:

- regulatory compliance;
- demonstrating due diligence;

- achieving best practice;
- identifying opportunities for competitive advantage;
- minimising ecological impacts; and
- minimising public health impacts.

Many of these goals are likely to overlap with the broader goals that organisations set themselves. By aligning the goals of environmental risk management efforts with the organisation's goals, it should also be possible to ensure that senior management actively supports the risk management process.

Because the management of environmental risks is a relatively new activity for many organisations, there is a general absence of criteria to assist in defining 'acceptable' risks. While some assessment techniques e.g. Quantified Risk Assessment, Public Health Risk Assessment, are reasonably well developed and have generally accepted criteria for the assessment of tolerability, most aspects of environmental risk assessment are not well developed and there is an absence of accepted criteria for the assessment of such risks. Furthermore, as environmental risk management can encompass a wide range of impacts and issues (e.g. ecological impacts, compliance issues, corporate image and public health impacts), it is frequently not clear how the relative importance of these different issues is to be assessed. In practice, organisations need to define their own risk acceptability criteria based on their own goals, objectives and targets, taking due account of relevant environmental legislation and environmental liability issues.

21.4   Identification

Risk identification is a structured process whereby risks and exposures are identified. One approach which could be adopted for risk identification is the development of a register of environmental effects and risks – similar to that required by the ISO 14001 (ISO, 2004). The advantage of such a register is that risks are recorded and can be reviewed in the future (e.g. for changes to legislation). This register also allows other parties to assess the reliability of the information, the comprehensiveness of the identification process and whether the organisation's objectives and scope have been adequately covered. Table 21.1 shows a suggested structure for such a register.

21.5   Analysis and assessment

Risk analysis is a systematic process to determine how often events may occur (*frequency*) and the magnitude of the likely *consequences*. Risk assessment involves comparing the predicted levels of risk against previously established risk criteria and deciding whether the predicted risk can be accepted. In addition, risk assessment should assist in the identification of priorities for action.

Table 21.1   Register of environmental effects

Activity	Identify the activities, operations, processes and procedures which impact, or could impact the environment. The range and extent of the activities considered depends on the scope and objectives of the risk assessment process but would probably include: (a) those activities, operations or products which release or have the potential to release anything to air, land or water; (b) operations or products which create any change in the environment (positive or negative), or have the potential to do so; (c) activities, operations or products that generate wastes.
Media	Identify the environmental medium affected (or potentially affected) by the activity. This would include air, water, noise, soil, waste, hazardous waste and raw materials usage.
Impact	Identify the effects or impacts of the activity. Consideration should be given, not only to the impacts of routine operations, but also the effects of abnormal or non-routine operations (e.g. start-up, maintenance) and incident or accident situations. The sources or causes of risks should be identified as part of this process.
Regulations	Identify all legislation, standards, licenses and codes of practice which apply to the activities, environmental media or impacts identified.
Other Requirements	Identify any other requirements which may be relevant to the activity or impacts under consideration. This could include corporate or stakeholder requirements identified as part of the context establishment.
Current Controls	Identify the current controls for dealing with the activity or impacts. The controls could include engineering controls, procedures, emergency response plans, monitoring programmes, reviews and training. The controls identified should include the controls in place for normal, abnormal and accident situations.

One of the major difficulties with environmental risk assessment is the wide range of consequences and impacts that may require consideration. In addition, there are significant gaps in our knowledge concerning the environmental and ecological impacts of pollution e.g. the dose-response of ecological systems to chemicals is very poorly understood (Sullivan and Hunt, 1999). As a result, a qualitative approach is often adopted for the analysis of environmental risks, involving the use of descriptive scales (e.g. high, medium, low) to the analysis of frequency and impacts, where each of these terms are clearly defined to ensure that different impacts are treated on a common basis. The use of qualitative analysis techniques does not necessarily preclude the use of more refined assessment techniques (e.g. Public Health Risk Assessment) at a later stage. Frequently, qualitative risk assessment is used to obtain a general indication of the level of risk and to assist in deciding whether more detailed studies are required.

Worked example: qualitative approach

Table 21.2 provides a relatively simple means of characterising the *likelihood* of a specific risk occurring, while Table 21.3 shows how the significance of environmental *consequences* may be assessed. The categorisations presented in Tables 21.2 and 21.3 are not intended to prescribe specific assessment frameworks (e.g. organisations may use more/fewer categories or different descriptors) but to illustrate how companies can consider and assess likelihood and consequences. For the assessment of likelihood, routine operations would be expected to fit into categories L4 or L5, while non-routine (abnormal) activities or operations would fit into categories L2 or L3. Rare (or very low frequency) events would be in category L1.

Environmental impacts are just one of the endpoints that need to be considered when assessing *consequences*. Similar tables may need to be developed for other endpoints (e.g. public attitudes, corporate reputation, legal compliance status). For public attitudes, the assessment could range: from issues that are of no identifiable concern to the public (C1: insignificant consequences); to issues that lead to intermittent complaints being received (C3: moderate consequences); and to issues likely to lead to significant negative press coverage (C5: catastrophic for the company's reputation). For facilities or operations that present significant risks to public health or the environment (e.g. petroleum refineries), this question

Table 21.2 Likelihood assessment

Likelihood category	Descriptor	Description
L5	Almost certain	The activity or operation occurs almost all the time
L4	Likely	The activity or operation occurs most of the time or very regularly
L3	Moderate	The activity or operation occurs some of the time
L2	Unlikely	The activity or operation could occur at some time
L1	Rare	The activity or operation may occur only in exceptional circumstances

Table 21.3   Consequence assessment (environmental impacts)

Consequence category	Environmental impact	Descriptor	Example(s)
C1	Insignificant	No off-site environmental impact identified	Ambient noise levels at site boundary are inaudible
C2	Minor	Some off-site impact which are considered acceptable	
C3	Moderate	Localised impact, or a small contribution to a significant regional or global issue.	Congestion at an intersection due to traffic to/ from the site; contribution to total photochemical smog pollutants in area
C4	Major	Major regional or local impact, which is reversible or is a major contribution to a significant regional or global issue.	Major spill of oil off-site; release of nutrients into local river system
C5	Catastrophic	Major regional or local impact which is irreversible	Destruction of a native animal habitat or river ecology

of public attitudes or perceptions may be particularly important as public concerns do not appear to conform to scientific or technical measures of risk (Kemp, 1996; Slovic, 1991). For example, activities with the potential for catastrophic consequences are generally regarded as being a greater risk and familiar risks e.g. those risks associated with driving a car are generally regarded as being less significant than unfamiliar risks. See Sullivan (2000) for a discussion of this issue and of the manner in which risks can be communicated to the public.

The overall *consequence* for an activity or operation is the maximum consequence category for any of the areas (e.g. environmental impact, public concern) considered. For example, an unauthorised release of oil that leads to minor pollution of a local river may be relatively minor in terms of the environmental impact and in terms of public attitudes towards the organisation (i.e. the incident may be considered as a minor (C2) incident in both of these consequence categories). However, the non-compliance with environmental legislation may be a major (C4) issue for the organisation.

Table 21.4 Defining environmental risk priorities

Likelihood	Consequences				
	C1	C2	C3	C4	C5
L5	Significant	Significant	High	High	High
L4	Moderate	Significant	Significant	High	High
L3	Low	Moderate	Significant	High	High
L2	Low	Low	Moderate	Significant	High
L1	Low	Low	Moderate	Significant	Significant

In such a situation, the overall consequence rating would be considered to be C4 (major).

When assessing *likelihoods* and *consequences*, it is necessary to consider the engineering and management controls which are in place to control the impact e.g. a spill of crude oil into a bunded area would not normally lead to an off-site impact (assuming that the bund floor is impermeable) and thus the environmental impact would be considered insignificant. In this context, there would be two scenarios to be considered (a) the rare (L1) event where the bund fails and there is a significant release of oil to the environment leading to the destruction of the local river ecology (C5), and (b) the unlikely event (L2) of an oil spill occurring where the bunding or other containment systems work as intended and there are no off-site environmental impacts (C1).

Having assessed *likelihoods* and *consequences*, organisations then need to assess the significance of the different *risks* e.g. an organisation may wish to categorise its risks as follows:

- high (requires immediate action by management);
- significant (requires management to develop an action plan within 6 months);
- moderate (requires management to develop action plan within 12 months); and
- low (where no short to medium actions required, however priority should be reviewed within the next 12 months).

One model for the relationship between likelihood and consequence and priority is outlined in Table 21.4.

Semi-quantitative approaches

For some organisations, the qualitative approach outlined above may not provide sufficient differentiation between risks and consequently a semi-quantitative approach may be adopted, where numerical values are assigned to the descriptive scales and mathematical formulae (e.g. *Risk = Frequency* multiplied by *Consequence*) are then used to estimate the level of risk. Using the example above of the oil spill into the bunded area, the risk of the spill occurring but not escaping from the bund could be scored as 2 (i.e. frequency of 2 multiplied by a consequence of 1), while the risk of the bund failing and leading to pollution of the local river could be scored as 5 (i.e. frequency of 1 multiplied by a consequence of 5). While such analysis may provide additional guidance to management on which risks need to be prioritised, quantitative approaches may create an impression of false precision that does not reflect the quality of the underlying data. Therefore, care is required to ensure that a spurious or misleading accuracy or importance is not assigned to the results of this type of analysis.

21.6   Treatment

Risk treatment involves identifying options for treating risks and evaluating the available treatment options while acknowledging that the resources available to manage risks are not infinite and that many risks may need to be accepted and tolerated or controlled rather than eliminated. The risk treatment strategies to be adopted will involve consideration of the costs and benefits of the available strategies as well as factors such as operations, resources, ease of implementation. The benefits need to be assessed in terms of the reduction achieved in the estimated level of risk.

The options for managing risks may be divided into the following categories (Standards Australia, 2004, pp. 20–22):

- *prevent or eliminate* the risk;
- *reduce* the risk through reducing either or both of the likelihood of occurrence or the consequences in the event of an incident;
- *transfer* the risk through, for example, purchasing insurance or outsourcing certain activities to third parties; and
- *accept* the risk. This last is a very important point as the organisation may decide that certain risks can be accepted or that the costs of addressing the risk may be too great relative to the benefits received.

For each risk, the options for managing the risks should be subject to a cost-benefit analysis to determine whether or not the action required to prevent or eliminate the effect can be justified. Such cost-benefit assessments should also include the financial benefits, such as reduced costs or improved productivity which may result from the change.

It is important to note that the purpose of risk management is not necessarily to eliminate risks but to ensure that risks are appropriately managed consistent with the organisation's objectives. For virtually all organisations, there will continue to be a need for emergency response plans to address residual risks.

Following the decision on the risk treatment strategies to be adopted, a treatment plan should be developed. The plan should detail:

- how the risk will be controlled?

Controls can include engineering (e.g. process modification, provision of physical controls), operational measures (e.g. procedures, training, supervision), emergency response systems (e.g. emergency response plan preparation, implementation and rehearsal) and strategic aspects (e.g. defining contractor and supplier responsibilities).

- what are the resources required?
- who is responsible?
- when must the strategy be implemented?
- how performance will be assessed?

The plan should include provision for the review, not only of the implementation of the risk reduction measures, but also the effectiveness of the strategies adopted with provision for corrective action in the event of failures of the plan.

21.7 Monitoring and review

Risk assessment and risk management are not static processes. Procedures should be implemented to ensure that the environmental risk management system is monitored and reviewed on a regular basis. The review process should not be confined to the implementation of the strategic plan but should also include:

- review of strategic context, objectives and criteria with particular attention focussed on the management of change (e.g. changes to legislation) and the integration of change into the risk management process;
- review of the risk register (at least annually) to ensure the register is current and complete. The register should be reviewed and updated whenever there are changes to any of the parameters which relate to the estimated level of risk. Such changes could include changes in regulatory requirements, process or procedural modifications and alterations in corporate requirements: and
- review of the risk analysis to ensure that the analysis and priorities assigned are still relevant to operations – whenever the risk register is updated or at least annually.

21.8   Documentation

All assumptions, data sources, calculations and methods used in the risk assessment and risk management process should be clearly recorded. This enables the risk management process to be audited and facilitates the ongoing monitoring and review of the risk management system. For environmental risk management, the documentation of the process is of particular importance given the increasing legislative requirements for organisations to demonstrate due diligence in environmental management. The documentation typically required will include organisation charts, procedures, emergency response plans, and environmental incident management guidelines covering reporting, recording, investigation and corrective action. In addition, suitable records will be required to demonstrate that the risk management system is effectively implemented. These may include tests of emergency response systems, reviews of the performance of the system, minutes of meetings where risk management issues are discussed.

21.9   Discussion and conclusion

In many organisations, environmental risk management is treated simply as a subset of general environmental management. In others, risk management is a stand-alone area of activity. In practice, it is for organisations to decide how best to manage environmental risks in the context of their operations and activities. The framework presented here is intended to be generally applicable, irrespective of whether environmental risk management is treated as a stand-alone activity, as part of overall environmental management systems or as part of risk management systems. Wherever possible, environmental risk management systems should be integrated with existing management systems and processes to minimise the costs of implementing the system and facilitate the implementation of the system.

One of the emerging trends in the risk management literature has been a growing focus on the business opportunities presented by better risk management (Standards Australia, 2004). Conceptually, this argument has much to recommend it. However, it may not fit particularly well with the common risk management priority for companies of the avoidance of loss (from accidents, spills, non-compliance with regulations, etc.). The decision on whether or not to include opportunities assessments in risk assessment is, ultimately, an organisation-specific issue. Two specific comments need to be made in this regard: organisations generally use other systems and processes to identify opportunities: and, as commonly practised, risk assessment and management processes are particularly suited to the avoidance of loss.

Ultimately, environmental risk assessment processes are about allowing companies and managers to make informed decisions on the management of risk, through providing a structured approach for assessing competing demands and deciding how inevitably restricted budgets are to be spent.

21.10 Questions

* Using your organisation's operations develop a register of environmental effects and impacts (Table 21.1). Undertake a risk assessment using Tables 21.2 and 21.3 to determine for each impact its likelihood and consequences and then assess its risk level or priority (Table 21.4). Suggest treatments for the management of the risks (Section 21.6).

22
Cleaner Production

Robert Staib

22.1 Introduction

In this chapter, we outline some of the reasons for and categories of cleaner production. We then describe a management approach to undertaking cleaner production assessment and implementation within organisations. We conclude with a case study describing a typical process flow for a coal-fired power station to encourage readers to prepare their own cleaner production analyses. The references provide a source of further information.

Many terms are used in the literature for this approach with subtle differences between each of them e.g. pollution prevention, clean technology, eco-efficiency and environmentally benign manufacturing (Overcash, 2002). Unless otherwise noted, we use the term cleaner production in a generic sense to mean the application of various process, techniques and tools to the task of reducing the environmental impact of manufacturing.

22.2 Reasons for cleaner production

Cleaner production was initially applied to the prevention of pollution and is an effective response to help an organisation comply with the legislation that has sought, increasingly over the last 50 years, to control emissions to land, air and water and to manage the increasing volume of waste produced by urban areas and industry (Overcash, 2002). It is a structured management approach to the analysis of the environmental impact of manufacturing. It requires involvement of the whole organisation and seeks to achieve both environmental and cost targets within a business framework.

This approach is needed to address the problems of unsustainable resource use (materials and energy), the use of non-renewable resources and increasing pollution (excessive land use for waste disposal and the release of toxic by-products). While compliance with environmental legislation and the desire to improve an organisation's environmental performance are important, one of the prime motivating forces for cleaner production is cost reduction (Cagno et al, 2005; Gutowski et al, 2005).

Other forces include pressure from the public, from the market, from stricter emission standards and increasingly legislative pressure for organisations to assume more responsibility for their emissions, wastes and products throughout their lifecycles e.g. product take-back or extended product liability (van Berkel et al, 1997).

Table 22.1 Evolving categories of cleaner production

Category	Component	Example
Pollution control	Disposal	Construction waste buried in a landfill site.
	Treatment	Process waste water treatment plant with disposal to a local creek.
Waste management	Energy recovery	Waste wood chips burnt to provide process steam.
	Reclamation and reuse	Demolished concrete crushed to produce base for new road.
	Off-site recovery	Off-cut from steel sheets from manufacture of drums returned to steel producer for re-smelting.
Cleaner production	On-site recycling	Process wastewater treated to return to manufacturing process.
	Source reduction	Product designed to use less packaging e.g. McDonalds replaced polystyrene with paper packaging for its hamburgers.
	Resource recovery	Car bodies stripped, materials separated, processed for reuse.
Cleaner technology	Product take-back	Dutch legislation requiring producers of computers and telephones to take back their products at the end of their lives.
	Dematerialising: Supply a human need with a service not a product	Producer owns product and controls use/reuse/recycling by leasing to user e.g. photocopiers.
Industrial Ecology	Cascading networks of industries and users sharing inputs and wastes	An eco-park where wastes from one industry provide feeds for another, combined waste is processed to supply process heat and electricity to the industries and transport costs reduced because of proximity of industries.

Sources: Gutowski et al (2005); Environment Protection Authority (2000b); Cagno et al (2005); Overcash (2002); Clift (1997); Hawken et al (2000, p. 134); Russo (1999, p. 359).

Table 22.1 shows an approximate categorisation of cleaner production approaches showing a progression from end-of-pipe pollution control techniques to management orientated and strategic planning approaches. Industrial ecology approaches are in their infancy while pollution control approaches are still common.

22.3   Cleaner production management

This section discusses the management processes associated with cleaner production (Figure 22.1). The reader may note that the cleaner production management processes are similar to some of the management processes described in other chapters.

Plan and organise

Before embarking upon a (or revising an existing) cleaner production strategy it is important to obtain senior management support and commitment, as a precursor to seeking approvals for the necessary resources of people and funds. For organisations commencing a new strategy it is advisable to commence with pilot projects and when successful extend to other areas of the organisation. Clear and agreed objectives established early will focus the organisation's attention on achievable outcomes (Environment Protection Authority, 2000b).

Initial Assessment

Prior to embarking upon detailed analyses, it is necessary to gather the technical and historical data to establish a good understanding of the organisation's products and its manufacturing and production process. This includes collecting existing and new data on manufacturing process inputs (energy and materials), product outputs, unwanted outputs (emissions to air, land, water), wastes and the direct and indirect costs of producing the emissions and wastes. A process flow diagram linked to a mass balance will assist in identifying all wastes and opportunities for eliminating them from the manufacturing process. Figure 22.2 is a simplified flow diagram of a coal-fired power station, which could be built upon as information is gathered. At this stage the types of tools or techniques that could be used to gather and analyse the data, can be investigated (van Berkel et al, 1997). Some are shown in Figure 22.1.

Cleaner production options

Much published information is now available to guide companies. Guidelines have been prepared for many industries that outline options available for cleaner production and this includes specific quantitative data on achievements of environmental and financial gains from cleaner production projects and programs across many industries. This data is published

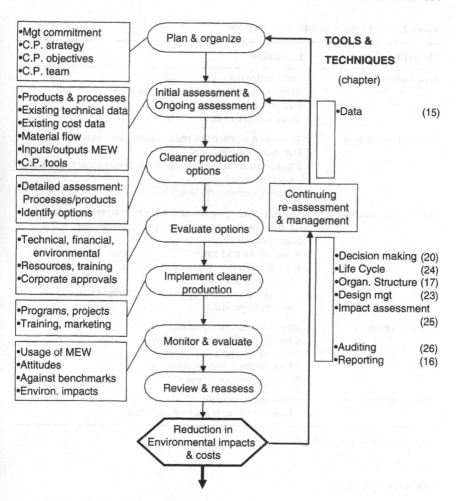

Figure 22.1 Cleaner Production (CP) management processes
Source: Based on Environment Protection Authority (2000b), Overcash (2002), van Berkel et al (1997). CP = Cleaner production; MEW = materials, energy, waste.

in hard copy and on the Internet by bodies such as the EPAs, United Nations and industry associations. Table 22.2 outlines a checklist of items to consider, when deciding on options for a cleaner production project.

It is important to identify a range of production areas, opportunities and projects for assessment and within a particular project identify several options for implementation. Initial assessment can assist in prioritising projects to identify which have the greatest potential to achieve initial success with the least expenditure on funds.

Table 22.2   Cleaner production options

Element	Examples
Raw materials	Reduce hazardous materials Avoid generation of hazardous wastes Buy purer materials Make material substitutions
Technology change	Modify equipment to reduce waste and emissions Use more automation Change process conditions: flow, temperature, pressure, resident times Improve energy efficiency
Better housekeeping	Change management and administrative procedures Change operating practices, provide training Segregate waste Avoid leaks and spills Introduce better production scheduling, handling and inventory practices Change accounting practices e.g. direct charging of wastes to products
Product changes	Improve quality standards Change product composition Improve product durability Make product substitutions Redesign products, use LCA
Reuse and recycling	Reuse waste on-site Make use of recycled raw materials

Source: United Nations (1996, 2004b).

Evaluate options

Cleaner production programs and projects will be subject to the same controls as other programs and projects and will involve most key departments within an organisation and will require justification before senior management will agree to their go-ahead. Therefore they will need to be subject to technical, financial assessment as well as environmental assessment. The environmental assessment will need to be rigorous to be able to stand alongside the other two traditional aspects. Because a cleaner production program could involve, over the medium term, a substantial change to the way an organisation does business, the organisational and cultural changes needed should be identified.

Organisational aspects

Three organisational aspects need to be addressed to avoid hampering the implementation of cleaner production (or pollution prevention) projects and programs. They are the need to: change organisation culture; have the right people in the right place; and navigate the organisation's politics. Failure to adequately address these aspects can affect important decisions-making points associated with identifying the cleaner production opportunities, specifying and agreeing solutions and implementing those solutions (Cebon, 1993). There are many case studies in the literature that provide insight into how to address these issues. We summarise one below that illustrates some of the issues that can affect success.

Case study: washing machines

Over a three-year period Vickers (2000) studied how a company making washing machines introduced and implemented a cleaner production project. The initiative was marketing-led with the objective of raising the profile of a new range of environmentally friendly washing machines. The company established an Environmental Affairs Committee, which was chaired by the Director Product Development and included representatives of the production engineering, marketing, distribution, product servicing and the safety and environment departments. A number of organisational issues (which most organisations typically face from time to time) restricted the success of the project. They included an immediate history of financial losses and cost cutting, industrial relations problems, loss of key people who were supporting the initiative and loss of commitment from certain executives.

Implementation

Implementation needs to be progressive going from pilot projects, multiple projects to a program involving large parts of the organisation. Marketing, education and training are important to the success of the program.

Management feedback loop

As with other management processes it is important to include the feedback loop of monitoring achievement, evaluating, reviewing and reassessing plans to produce a culture of continuing improvement. Accounting and technical information systems will need to be established to record performance and provide sufficient information to be able to review progress and establish new priorities and programs.

Cost reduction

Because one of the prime motivating forces for cleaner production is cost reduction (Cagno et al, 2005; Gutowski et al, 2005), gathering cost information and using it to justify a cleaner production project will be an important part of obtaining organisational approval of a project. Having reliable cost information will enable better environmental decisions to be made. There are many ways to gather and to categorise cost information but the following list will suffice to illustrate the point:

Internal
- direct cost of materials and energy embodied in emissions and waste;
- indirect cost of producing waste e.g. extra organisational overheads, utilisation of extra capacity of staff and manufacturing equipment;
- cost of treatment and disposal of waste and emission fees;

External
- life cycle costs of using and disposing of products; and
- environmental clean-up costs from emissions created in the production and use of products.

Tools and techniques

There are many different tools and techniques that can be used to assist the cleaner production process. Some are shown in Figure 22.1 opposite the relevant cleaner production process and include a reference to a chapter of this book where they are described e.g. life cycle assessment (Chapter 24), environmental design management (Chapter 23) and auditing (Chapter 26). They can be supported by checklists and spread sheets for recording and producing material and energy balances (Environment Protection Authority, 2000b).

22.4 Summary and implications for business organisations

We have used the term *cleaner production* in a generic sense to mean the application of various management processes, techniques and tools to the task of reducing the environmental impact of manufacturing. Some related approaches have been listed and briefly described e.g. pollution prevention, clean technology and industrial ecology.

In Figure 22.1, we describe a management approach to the task of undertaking cleaner production assessments and projects within organisations. Table 22.2 can be used to assist in the selection of cleaner production projects. We refer the reader to the significant literature on specific industries, much of it on the Internet that provides guidance and quantified case studies. Many industries are using an incremental approach to cleaner production though it lends itself to a more revolutionary approach (van Berkel, 2000).

There are many reasons why organisations should seek to make their production cleaner including: pressure from the public and the market; requirements of stricter emission standards; and increasing legislative pressure for organisations to assume more responsibility for their emissions, wastes and products throughout their lifecycles. While compliance with environmental legislation and the desire to improve an organisation's environmental performance are important, one of the prime motivating forces for cleaner production is cost reduction.

Case study: coal fired power station

Coal fired power stations produce a significant amount of greenhouse gases and the task to reduce these gases and to reduce the environmental impact of global warming is large but some electricity producers are already tackling this significant environmental management issue. Figure 22.2 shows a preliminary flow chart of the production of electricity from coal with the major inputs, the internal processes and the outputs in terms of electricity and wastes. This is a first step in developing more detailed flow charts, mass balances and identifying opportunities for *cleaner production*. While the major impact is the emission of carbon dioxide, it is probably the hardest to treat but there are many opportunities to achieve early gains with control of dust from each operation, reuse of fly ash in concrete manufacture and treatment and reuse of waste water.

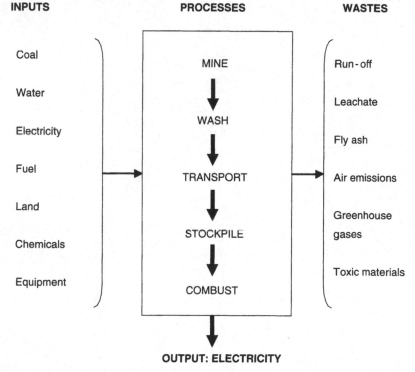

INPUTS

Coal

Water

Electricity

Fuel

Land

Chemicals

Equipment

PROCESSES

MINE

WASH

TRANSPORT

STOCKPILE

COMBUST

WASTES

Run-off

Leachate

Fly ash

Air emissions

Greenhouse gases

Toxic materials

OUTPUT: ELECTRICITY

Figure 22.2 Flow chart for production of electricity from coal
Source: Based on (Pagan et al, 2003; Delta Electricity, 2003).

22.5 Further reading

The World Bank (1999); Graedel (2000); Asian Development Bank (2002).

22.6 Questions

- Of the many of the process models shown in this book the model for *cleaner production* (Figure 22.1) could possibly be seen as the master that links all others e.g. life cycle assessment (Chapter 24), design management (Chapter 23), decision making (Chapter 20) and auditing (Chapter 26). Is cleaner production the 'theory of everything' for environmental management within an organisation?
- Several of the authors quoted in this chapter indicate that one of the prime motivating forces for *cleaner production* is cost reduction. Does this mean that the environmental objectives of cleaner production are being lost to the ever present need of business organisation for financial survival?

23
Environmental Design Management

Robert Staib

23.1 Introduction

This chapter discusses environmental design management and its role in the delivery of products and infrastructure to help reduce the impact they make on the environment and how, if applied progressively, can help bring about a more sustainable use of resources. Design management is the management of the design process that is carried out typically by architects, engineers, industrial designers, inventors and others who design products for use and consumption by the community or who design infrastructure to serve the broader needs of community.

The traditional design management process embodies environmental impacts into products or infrastructure that can manifest themselves throughout their economic life or life cycle. Environmental design management seeks to introduce consideration of the environmental impacts into the traditional design management process. This chapter draws on some of the ideas of traditional engineering and architectural design management and the emerging ideas on environmental design management (Staib, 2003), Design for Environment (Environment Australia, 2001, Yarwood and Eagan, 1998) and product life cycle assessment (Keoleian et al, 1994).

23.2 Product life cycle

Products are a part of the built environment and are those things produced by humans as distinct from the humans themselves and the natural environment. They range from the short-lived products like the daily newspaper of 1 day to long lived infrastructure of up to 70 years e.g. a railway system.

They include: consumer products (microwave ovens, cars, houses); industrial products (manufacturing equipment, buildings); and infrastructure (roads, sewerage, water systems and power supply systems). A product could include a pop concert and all the resources necessary to stage it.

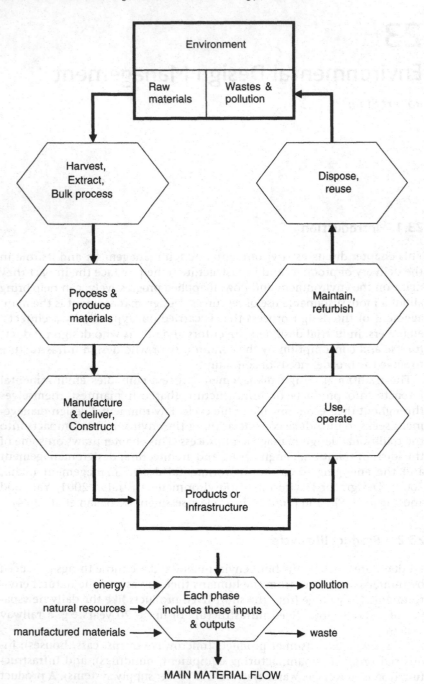

Figure 23.1 Product life cycle – main material flows
Source: Based on Environment Australia (2001); Keoleian et al (1994); Staib (2003); Yarwood and Eagan (1998).

Figure 23.1 illustrates the concept of product life cycle by showing the flow of the main materials needed to produce and sustain a product. It shows 6 broad phases in a product's life with each adding value to, or maintaining the value of, the main material flow and having secondary inputs of energy, natural resources, manufactured materials and secondary outputs of pollution and waste. Labour, intellectual energy, cost and revenue could perhaps be included.

The life cycle is an important concept in considering and accounting for all possible environmental impacts linked to our use of products and infrastructure. It also forms the base from which various environmental management tools derive their systematic nature e.g. life cycle assessment and life cycle costing, environmental impact assessment, cleaner production and environmental decision making (or choosing between alternatives). See the other chapters in Part IV of this book for a discussion of some of these environmental management tools.

23.3 Design management process

Figure 23.2 (in its centre stream) shows the traditional design management process with 7 sub-processes. The left hand boxes show how environmental considerations (shown bolded) can be added to the sub-processes and the right hand stream shows the ongoing corporate systems that are needed to support the product development and the design management process. These systems need to integrate environmental issues into mainstream business issues to provide a supportive culture with ongoing market analysis, research and development, an EMS, environmental research and operations research.

The design management process model is shown simplified as a linear model with 7 sub-processes but would in practice contain many feedback loops. The model could apply to each of the 6 phases in Figure 23.1 where design is required. The Design Manager could use one or more of the many environmental design tools becoming available e.g. life cycle assessment and life cycle costing, environmental impact assessment, cleaner production and environmental decision making (or choosing between alternatives) as well as being assisted by the organisation's Environmental Manager.

The advantage of using the environmental design management model of Figure 23.2 is that it matches and integrates with the traditional design management approach. This is important because environment is only one, albeit an increasingly important aspect of design. Figure 23.3 shows a typical arrangement of interlocking issues that a designer needs to consider, in varying degrees depending upon the product. Issues include customer requirements and function as the core issues with supplementary ones including: industry standards and safety, environmental issues, weight and size and appearance, ease of manufacture or construction, cost, revenue and timing, and ease of maintenance and operation.

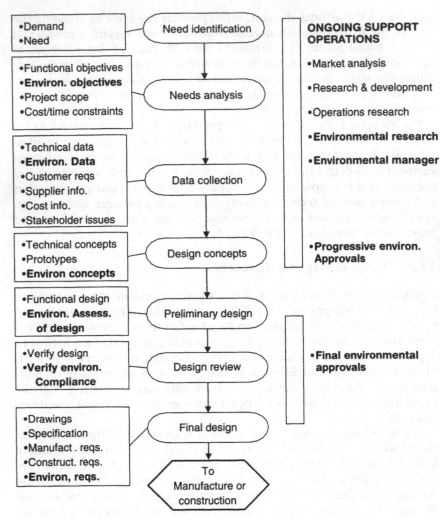

Figure 23.2 Design management process with an environmental focus
Source: Based on Environment Australia (2001); Keoleian et al (1994); Staib (2003); Yarwood and Eagan (1998).

Needs identification

At the first sub-process Needs Identification (Figure 23.2), a client (e.g. an external client for infrastructure development or an internal client for product development) makes decisions that will effect the environment. The decision by Sydney Australia to make a bid for the 2000 Olympic Games immediately embodied environmental impacts, which became even more significant when Sydney was successful (OCA, 1996; 1997).

Needs analysis

Design managers may become involved on a product development or a project when a Needs Analysis is undertaken and functional objectives are set. It is important at this time to also set environmental objectives. For an organisation with an environmental policy of continual improvement, this may in part come about automatically but it is beneficial to look at each individual product or project and set specific environmental objectives. 'The earlier that the design team considers the environmental factors the greater is the potential to achieve environmental benefits and cost reductions' (Yarwood and Eagan, 1998).

Data collection and baseline studies

The product development or project scoping should include generating and collecting environmental data. For product oriented organisations much could be available through market research and in-house environmental research (Toyota, 2004) but for organisations contemplating a new product or developing a one-off infrastructure project this may be a separate task (Staib, 2003). If the environmental data (baseline conditions, potential environmental impacts, life cycle properties of materials, manufacturing processes and the use and disposal of similar products) is not generated at the start of design along with the engineering and marketing data it can be bypassed or glossed over in the imperative (brought about by time and cost constraints) to complete the design and start manufacture or construction.

For designers the data needs to be in a form they can use i.e. it needs to be supported by a scientific and engineering rigour. There is no point in the designer specifying recycled material for the current project if there is no assurance of supply or no system to guarantee that the material will have consistent engineering properties and quality.

Design concepts

Development of design concepts marks the point where a large part of the creative input occurs as the designer strives to address the customer or client requirements and to produce a design that will function in the manner required e.g. a building that meets the total rental floor area needed, has clear spans with minimum columns and a landmark façade. This is the time to structure the building for use of natural light, make arrangements for reuse of waste water and collection of rain water, identify opportunities for recycled or low impact materials that support the environmental objectives previously set and agreed.

Preliminary design – technical and environmental

The integration of environmental issues or components into the design can now take place based on the concepts established. While the Design Manager may be 'green' it is important to have a regular independent input from the organisation's Environmental Manager or environmental staff.

Progressive environmental assessment of design

It is important that those parties charged with the approval of environmental design matters including statutory authorities for infrastructure projects, emission authorities for vehicles or corporate environmental departments have an opportunity to progressively review the developing design to make final approval easier but also to provide input to improve the environmental performance of the final product or infrastructure.

Design review

Independent review of design is a mandatory process under organisational quality assurance systems especially if the organisation is accredited to the standard ISO 9001. This approach is also important from an environmental perspective to identify whether the environmental objectives have been incorporated into the final product or have been lost or watered down in the face of compromises with the other design issues shown in Figure 23.3.

Final design

The process of moving from preliminary to final design can sometimes be long and arduous as the needs and comments from all parties in an organisation are addressed. Acceptance of the final design is important from an environmental perspective because in the final rush to complete a design

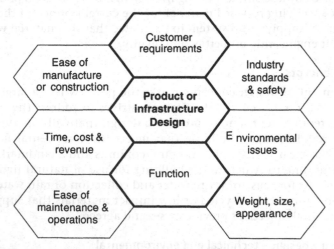

Figure 23.3 Design issues
Source: Based on Environment Australia (2001); Keoleian et al (1994); Staib (2003); Yarwood and Eagan (1998).

and make the compromises between competing claims important environmental features can be lost. If the stakeholders have been involved throughout the design process there is more likelihood that final acceptance will be straightforward. This could include final environmental approvals and formal statutory approvals.

Scope of environmental design management

Environmental design management can be applied over the six phases of a product's life shown in Figure 23.1. It can be applied to any number of products and types of infrastructure e.g. consumer products: houses, furniture, motor vehicles, newspapers, wine; industry products: buildings, equipment, metals, raw materials; infrastructure: sewerage, water systems, power stations and distribution systems, roads, railways, hospitals; and services or tertiary industry: concerts, sporting events, office environments. It is a systematic way to ensure all environmental issues are identified during the traditional design process and best affordable environmental practices are adopted.

23.4   Change from traditional design management

Design managers need to address environmental issues that come from a number of sources some within their control and some imposed upon them. Those outside their control can include: new or impending legislation, internal marketing initiatives, corporate research and development, guidelines from industry peak groups, world wide trends and international agreements, pressure from stakeholders (suppliers, clients, customers, owners, shareholders), government, community groups and green groups.

These increasing environmental pressures on organisations (described in Part I of this book) can act on designer managers and require them to change their traditional way of designing products. There is a growing body of literature explaining how to go about it with many examples of significant achievements (Hawken et al, 2000; Staib, 2003).

Figure 23.4 is a simple framework that shows reasons why designers might change their traditional methods to embrace more environmentally responsive design. At the first level is *mandatory* change. We change because we must e.g. because new legislation has been enacted. This is the basic compliance level.

The second level is *proactive* change. We need to change because of expected changes to legislation, government policies and markets. For the organisation to stay in business it needs to know what new legislation or markets will be in place when some of its strategic and mid term decisions are implemented.

The third level is *ethical* change. We change because of internal organisational initiatives and these could be driven by a changing organisation

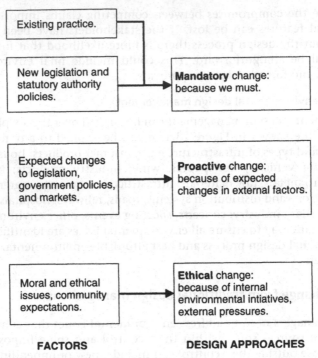

FACTORS   **DESIGN APPROACHES**

Figure 23.4 Need for change in design management
Source: Based on Welford (2000).

culture supported by moral and ethical issues or be driven by changing community expectations.

These last two levels can take an organisation beyond basic compliance.

Design managers

There is a need for design managers to change the way they manage design, to become more environmentally aware and to adopt better environmental design management processes because if they don't it may well be forced upon them.

This will involve a number of changes many of which are already underway. It will include establishment of quantitative environmental objectives along side the traditional engineering functional objectives, undertaking baseline environmental studies to provide the necessary environmental data to supplement the engineering data, development of quality standards for new environmentally friendly materials and achieving progressive liaison and approval with the various statutory authorities and stakeholders as design concepts are developed (Figure 23.2).

This is tantamount to integrating environmental impact assessment and mitigation into the design not just the minimum for statutory approval, but that extra amount to meet moral, ethical and community expectations and finally to take the design process towards the delivery of an environmentally sustainable product or infrastructure.

Although design managers do not have control over all the environmental issues that impact their designs, they can and should attempt to influence them. Design managers need to take a proactive role in establishing the product and project environmental objectives because the earlier an environmental impact is identified and mitigative features developed the greater the mitigation (Environment Australia, 2001; Staib, 1999; Staib, 2003; Yarwood and Eagan, 1998).

Design managers should be an important source of environmental design and impact data for their product or project. This should cover: the materials they specify; the energy embodied by their design (in the manufacture, construction, the on-going use of the product and the operations and maintenance of the infrastructure); the heritage they impact; and the environmental damage that results from their design decisions (or the decisions of others e.g. the client). They need to be aware of the new and emerging environmental assessment techniques (e.g. LCA, cleaner production) and the opportunities to influence change for better environmental outcomes.

Design managers should adopt the emerging practice of environmental design management and be proactive in making environmental design management a part of their profession, not leave it to others (Staib, 1999). Engineering and architectural design managers are well placed to take up this challenge. They are used to integrating the work of many disciplines and specialists into designs and environment is becoming one of the key disciplines and is likely to become even more important in the future (Hawken et al, 2000).

This places greater environmental responsibility on design managers resulting in the need for them to become more environmentally conscious and to acquire better environmental data and new environmental design management skills. Along the way they may develop a fresh outlook on the world. As many architects found during the Sydney 2000 Olympics, environmental design added a third dimension to the old adage of 'form and function' (Dinham, 1998).

23.5 Design for the environment

There are many other ways of visualising the design management process and how environmental aspects might be introduced to those presented in this chapter. Design for environment (DfE) (or ecodesign) has emerged as an approach to considering environmental aspects during design (Environment Australia, 2001). It makes use of a life cycle concept similar to Figure 23.2 by requiring designers to consider each phase. It is both a philosophy of and an approach to environmental design. Its approach consists of identifying a number of design related tasks and utilising a number of checklists.

The design related *tasks* are: analyse opportunities; promote DfE in the organisation; set goals and identify strategies; apply DfE tools; develop product; market product; and evaluate product. It is an alternative way of visualising some of the design process to that presented in Figure 23.2 and in this chapter. Understanding DfE is important because much literature is written around its framework.

Its framework incorporates the use of a range of environmental tools, as does the Figure 23.2 model though many of the environmental tools can stand-alone and do not need a strict framework. Some forms of DfE use a spreadsheet or matrix that can be used to decide between alternatives or to assess if a product reaches a certain set environmental standard. The matrix has as its rows the *tasks* (listed in the preceding paragraph) and as its columns: material use, energy use and waste (subdivided into solid, liquid and gaseous). Each element of the matrix is scored on a scale to assess the environmental performance of a product (Yarwood and Eagan, 1998).

23.6 Design management and other management approaches

The design management process would normally be project managed to achieve delivery of the design of the product or the infrastructure to the objectives for scope, quality, cost, time, environment and occupational health and safety (Staib, 1999).

Environmental design management can also be introduced into other traditional approaches used by design managers e.g. the Total Quality Management approach (Persico, 1992) which uses cross departmental teams to identify and instigate change and to achieve a quality approach for the whole organisation not just the quality of the produced or constructed product; and Value Analysis (or value engineering) which is an approach to cost reduction which eliminates unnecessary costs (Mudge, 1971; Thurwachter, 2000).

23.7 Conclusion and summary

We have presented Environmental Design Management as a management model that integrates environmental considerations into traditional design processes. 'Design for Environment' is an emerging philosophical approach that sets a framework for considering environmental issues and some proponents may suggest that it should replace traditional design. The author though suggests that it can transform rather than replace because, after all, environmental issues are one of the many issues a designer must

---

Case study: change of design objectives

Damage to Aboriginal archaeological sites is a consequence of construction of urban infrastructure in the expanding city of Sydney, Australia. Over a period of 12 years of building water infrastructure (sewerage, water and stormwater drainage works) for new suburbs in the north-western outskirts of Sydney, the design approach has evolved from one of identification of impacts and obtaining statutory approvals for those impacts to one of setting avoidance objectives and achieving them.

During Stage 1 (1990 to 1994) the archaeological sites to be impacted by the infrastructure construction were excavated in accordance with New South Wales National Parks and Wildlife Service's requirements. While this resulted in the gathering of archaeological knowledge, the result was the destruction of approximately 80% of the sites by the infrastructure construction.

By Stage 3 (2002 to 2004) the design management approach had evolved to a more environmentally responsive approach. Between the client, the designers and the statutory authorities a quantitative environmental objective evolved that required the avoidance of all aboriginal archaeological sites. This was a significant change from the Stage 1 qualitative objective of 'avoid damage if possible'. The design objective for Stage 3 was achieved by changing the way the design of the infrastructure was managed. This required new ways of thinking on behalf of the design managers and a willingness on the side of the client to accept potential new costs. The outcome was the protection of 70% of the sites with most of the impacted sites (30%) being sites of low significance. A similar approach was adopted to avoiding the native flora sites.

To support this changed approach to design management better baseline studies of existing archaeology, flora and fauna and progressive liaison with statutory authorities were required. (Staib, 2002; Staib, 2003a).

---

deal with – albeit an increasingly more important one as communities strive for sustainability.

There are a variety of environmental tools that a design manager now has available to 'green' his/her design and these need to be used as appropriate for the particular product or infrastructure and the particular phase of the design involved. Notwithstanding these new tools it is important not to disregard the tried and proven design tools like value engineering (Mudge, 1971) but to introduce a strong environmental flavour into them.

23.8 Questions

- How far can design managers go in achieving sustainable designs within the product cycle when they have no control or influence over many factors and issues?
- Select a particular product and assess the short-term targets for environmental design management e.g. up to 3 years, medium term targets 5 to 10 years and the long term say 10 to 50 years.

24
Life Cycle Assessment
Maria Atkinson

24.1 Introduction

This chapter describes the Life Cycle Assessment (LCA) technique used for assessing and comparing the environmental impact of products and materials. It discusses when and where it should be used, the reasons for using it and outlines the steps in undertaking an LCA. Strengths and weaknesses are discussed. The references lead the reader to important sources for software programs and data sources.

24.2 Background of LCA

Formalised in the 1990s, LCA is the assessment of the environmental impact of the manufacture and use of a product over its life from raw material extraction to disposal at the end of its life (Figure 23.1). ISO developed a methodology and framework for LCA in 1993 titled 'Life Cycle Assessment – Principles and Framework'. The LCA methodology has now been standardised by the International Organization for Standardization (ISO, 1997; 1998). Many countries have adopted this standard. As Klaus Töpfer, United Nations Environment Program Executive Director said in 2002: 'The impacts of all life cycle stages need to be considered comprehensively when taking informed decisions on production and consumption patterns'.

LCA requires the collection of extensive data and detailed evaluation of this data. Computer software programs have been developed to simplify data entry, evaluation and comparison of findings and worldwide databases are being developed to gather and store data on materials, processes and their environmental impact. LCA was originally developed for use by manufacturers considering options for product development e.g. it was used in the late 1960s in the United States by Coca Cola to compare the environmental impact of glass and plastic bottles (BuildingGreen Inc., 2002). In 1989, the Society for Environmental

Toxicology and Chemistry (SETAC) established itself as a coordinating organisation for the process and findings and it continues as a resource for publications on the ongoing development and use of LCA (Consoli et al, 1993).

Reasons for LCA

LCA has been regarded as a quantitative analytical tool; however, in recent years there is increasing interest in its role in decision making. To understand the environmental impacts of a product, process or service, it is important to note all the stages and impacts throughout their life cycles. LCA can aid decision makers in selecting the product or process that results in the least impact to the environment. LCA information can be used with other considerations and factors such as human health impact, cost and performance data.

'LCA data identifies the transfer of environmental impacts from one media to another (e.g. eliminating air emissions by creating a wastewater effluent instead) and/or from one life cycle stage to another (e.g. from use and reuse of the product to the raw material acquisition phase). If an LCA was not performed, the transfer might not be recognised and properly included in the analysis because it is outside the typical scope and focus of product selection processes' (USA EPA, 2001).

24.3 LCA processes and terms

LCA is an established scientific technique that involves four phases:

* defining the *goal* and scope;
* compiling an *inventory* of relevant inputs and outputs of a product system;
* evaluating the potential environmental *impacts* associated with those inputs and outputs; and
* *interpreting* the results of the inventory analysis and impact assessment phases in relation to the goal of the study.

Key terms used are: *environmental aspects* – those elements, activities, products or services than can interact with the environment based upon the criteria of significance presented in the study; *significant environmental aspect* – those environmental aspects that have, or can potentially have, a substantial positive or negative impact on the environment; *environmental impacts* – any change to the environment, whether adverse or beneficial, wholly or partially resulting from an organisation's activities, products or services; and *environmental significance criteria* – assessment criteria which may be set as aspects regulated by the law.

Table 24.1   Environmental aspects and impacts

Activity, product, service	Environmental aspect	Environmental impact
Coal fired power station: boiler operations	Air emissions from coal combustion	Air pollution, e.g. carbon dioxide, sulphur dioxide
	Coal consumption	Natural resource depletion (coal)
	Boiler blowdown & water discharge	Water pollution, e.g. temperature increase, oil
	Water consumption	Natural resource depletion (clean water)

Source: Based on Delta Electricity (2003); Woosley (2001).

In identifying *aspects* one looks at inputs such as raw materials, electricity, water use and outputs such as finished product, exhaust, waste products. A simple example of *aspects* and *impacts* is shown in Table 24.1.

24.4   Phases in LCA

LCAs can be conducted on a variety of elements, products, materials and processes, however, for ease of explanation the following relates to conducting a LCA on a particular product. The four phases in an LCA process are described in Section 24.3 and shown in Figure 24.1.

Goal definition and scope is the most critical and requires the assessors to clearly establish their objectives and the method and process to achieve it. This requires the assessor to document a description of the product, the scope of the study and the required reliability of the findings and the criteria for assessment.

Inventory analysis requires a systematic description of all the processes in a products' life cycle (Figure 23.1) and then collection of data on each process. The output is an inventory (often tabulated) of all raw material depletion and pollution (emissions or impacts).

Life Cycle Impact Assessment (LCIA) starts with sorting of all the raw material depletions, emissions and impacts according to their environmental effect. The calculations require the addition of indicators of different environmental effects e.g. an impact on air quality, an impact on water quality or a depletion of old growth forests, to produce a single overall score.

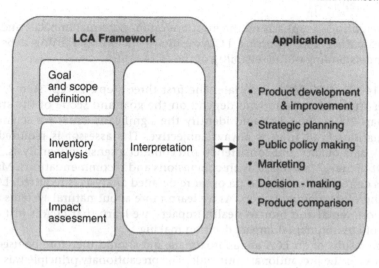

Figure 24.1 Phases of an LCA
Source: Based upon the figure from ISO (1997).

24.5 Steps in LCIA

The LCIA is the evaluation of potential human health and environmental impacts of the environmental resources used and emissions released as identified during the Inventory Analysis. Characterisation factors applied in the LCIA stage assist in calculating the impacts each environmental impact has on issues such as global warming. This requires value judgments, e.g. air emissions in one location could be of relatively higher concern than the same emission level in another location with better air quality. An LCIA may require seven steps:

1. Selection and definition of impact categories – identifying relevant environmental impact categories e.g. global warming, acidification, terrestrial toxicity;
2. Classification – assigning life cycle impacts results to the impact categories e.g. classifying carbon dioxide emissions to global warming;
3. Characterisation – modelling life cycle impacts within impact categories using science-based conversion factors e.g. modeling the potential impact of carbon dioxide and methane emissions on global warming;
4. Normalisation – expressing potential impacts in ways that they can be compared e.g. comparing the global warming impacts of carbon dioxide and methane;
5. Grouping – sorting or ranking the indicators (e.g. sorting the indicators by location: local, regional, and global);

6. Weighting – emphasising the most important potential impacts; and
7. Evaluating and reporting LCIA results – including gaining a better understanding of the reliability of the LCIA results.

ISO 14042 (ISO, 2000) mandates the first three steps. Except step 7, the other steps are optional and depend on the goal and scope of the study (Vehar, 2001). Step 4 should identify the significant issues for scientific evaluation. Evaluation is always subjective. The assessor is required to check data quality and consistency and conduct a sensitivity analysis. The result of Step 7 is a report with conclusions and recommendations. Many LCAs conclude that more data or more detailed analysis is required. LCAs are therefore moving targets. As we learn more about natural systems and environmental and human health impacts, we learn that LCA is just one tool of environmental impact decision making.

The results of an LCA are approximate and should therefore be used in context of the precautionary principle. The precautionary principle was first recognised in the World Charter for Nature and adopted by the UN General Assembly in 1982. It has been incorporated into various international conventions on the protection of the environment. (European Union, 2000) and was enshrined at the 1992 Rio Conference on the Environment and Development as Principle 15 of the Rio Declaration: 'In order to protect the environment, the precautionary approach shall be widely applied by States according to their capability. Where there are threats of serious or irreversible damage, lack of full scientific certainty shall not be used as a reason for postponing cost-effective measures to prevent environmental degradation'. The precautionary principle should not be confused with the element of caution that scientists apply in their assessment of scientific data.

24.6 Strengths

LCA is now a global tool for the assessment of global environmental impacts of products and materials. It can inform decision-makers by providing information which is often unconsidered. Initiated by the 2003 World Summit on Sustainable Development, three new United Nations environment programs have been established to promote LCA: Life Cycle Impact *Assessment* program to increase the quality and sharing of life cycle indicators; Life Cycle *Management* program to promote awareness and improve the skills of the assessors and decision-makers; and Life Cycle *Inventory* program to improve global access to transparent, high quality life cycle data.

24.7 Weaknesses

Performing LCAs can be resource and time intensive. Obtaining data can be difficult and the availability and accuracy of data can greatly impact the accuracy of final results and conclusions. LCA is subjective and often restricted to a limited number of environmental categories such as energy consumption and greenhouse gas emissions (See Case Study). LCA uses approximations, often simplifies the interconnections between natural systems and uses aggregated loadings (See Questions).

One of the biggest criticisms of LCA is that the findings and conclusions are often written without clear disclosure of the uncertainty or application of the precautionary principle. Some industry commissioned LCAs are used as a marketing tool to convince product procurers or stakeholders of the environmental benefits of a particular product or material, e.g. in the building industry, architects and interior designers often have no scientific training or experience to evaluate the claims.

'Life Cycle Assessment cannot provide a truly comprehensive and all-encompassing assessment' (Todd and Curran, 1999) because LCA does not directly measure the actual environmental impact, predict environmental effects, or take into account technical performance, costs, or political and social acceptance. LCA should be used in conjunction with these other parameters (US EPA, 2001).

LCA should not be therefore considered as a definitive answer. The guiding principles of ecologically sustainable development are rarely adequately addressed in the pure world of data collection and isolated system thinking attributed to most LCAs (Cowell et al, 2002).

One of the most recent articles on LCA 'Life-Cycle Assessment for Buildings: Seeking the Holy Grail' provides this fitting summary: 'If all this makes you think LCA must be an impossible challenge, you're right – the perfect LCA has never been performed. But many solutions are being pursued, addressing all aspects of the problem. Some of these are making the results of LCA studies more useful and accessible today, while others are in the works for the near or not-so-near future' (BuildingGreen Inc., 2002).

24.8 Summary

In this chapter we have outlined the reasons for and steps in undertaking an LCA. Some of the strengths and weaknesses have been discussed. It is a technique that is being used worldwide and a large amount of data is being generated to help make the LCA outputs more realistic and accurate. It is an important tool for business organisations to help them make decisions on their products and processes and for designers to help them choose materials and systems that have a lower environmental impact.

For further detail guidance in undertaking a LCA the reader is referred to the ISO 14040 Codes and the various guides developed by government authorities and industry associations in many countries, many available on the Internet (Section 24.9).

Case study: alternative road designs

Figure 24.2 shows a simplified flowchart for an LCA prepared to decide on which of two types of road construction for the Sydney 2000 Olympics had the least environmental impact: asphaltic concrete (or bitumen) or reinforced concrete. It covered extraction and processing of materials, construction, operation, maintenance and demolition of the two types of road construction over a life of 40 years.

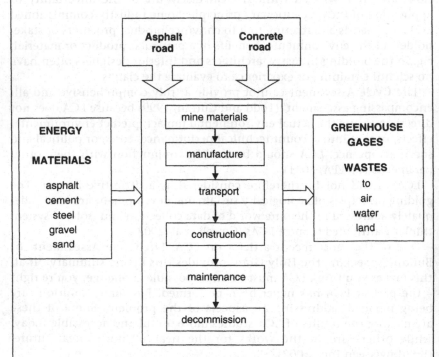

Figure 24.2   LCA inputs, processes, outputs for alternative road designs

Table 24.2 shows the results. The LCA suffers from many of the issues raised in this chapter of over simplification, not calculating the real impact on the environment and focussing on two broad environmental indicators.

Table 24.2   Example of a simple LCA for road construction alternatives

LCA Indicator (average over a 40 year life cycle)	Units	For asphalt road	For concrete road
Energy	MJ/m^2/year	14	35
Greenhouse	Kg CO^2/m^2/year	1.4	4.1

Source: Beer (1996).

Case study: alternative road designs – *continued*

Nevertheless it illustrates how LCA can be used in decision making with the concrete road having a significantly greater impact for both indicators. If one indicator had favoured concrete and the other asphalt then the decision making would not have been so easy.

24.9 Further reading

Text: Curran and Young (1996), Evans and Ross (1998), UNEP (2002). Web sites for LCA tools and data: EcoSpecifier http://www.ecospecifier.org/; LCAid™ http://asset.gov.com.au/dataweb/lcaid/; LCADesign http://www.construction-innovation.info/; and LISA http://www.lisa.au.com.

24.10 Questions

• One of the most subjective steps in LCIA is normalisation or weighting of the different impacts to enable comparison between options. Select a published LCA on a particular product and determine if using different weightings would produce a completely different outcome. See Bengisson and Steen (2000) for some ideas.

25
Environmental Impact Assessment
Robert Staib

25.1 Introduction

This chapter outlines the processes involved in environmental impact assessment (EIA) from an organisational point of view. EIA consists of the preparation of an Environmental Impact Statement (EIS) by the proponent of a development or project followed by a review by the public and government authorities (Thomas, 1998) or competent authority (European Union, 2001). After review the development may be approved or rejected or approved with conditions.

An EIS is normally required for large developments or smaller ones that are particularly controversial or have a significant impact. Other developments may require less impact assessment here referred to as an environmental review but having other names in different countries.

Organisations should have an EIA process for all their developments regardless of size and this should be an integral part of an EMS (Thomas, 1998) – with formal environmental approvals being provided by the organisations' independent environmental section (Chapter 12 on EMS).

25.2 When EIA is required

There is an extensive literature on the EIA process and a large number of EISs have been produced in many countries around the world. Many are available for the public to review. Developments include private and government projects from urban infrastructure and buildings through industrial facilities and factories to extensive agricultural industries. Some countries apply the EIA process to government policies and urban planning schemes and use the broader techniques of Strategic Environmental Assessment (SEA) and Cumulative Impact Assessment (CIA) (Court et al, 1996). SEA and CIA are mostly used by governments (Section 6.3) though in larger EISs organisations may be required to undertake a limited amount of cumulative impact assessment.

25.3 The EIA process

The EIS process is described in the relevant planning legislation (e.g. USA: National Environmental Policy Act 1970, European Union: EIA Directive 85/337/EEC, Australia: Environmental Planning and Biodiversity Conservation Act 1999, New South Wales, Australia: Environmental Planning and Assessment Act 1979). Figure 25.1 shows the main features of the EIA process.

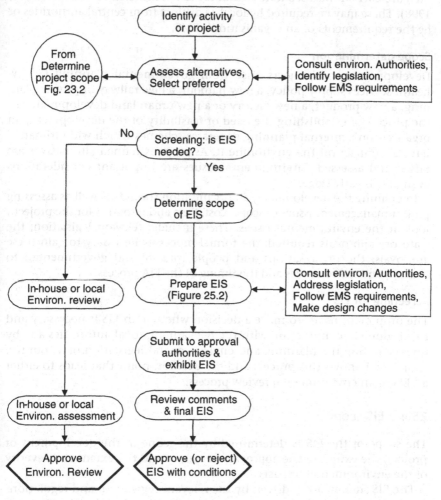

Figure 25.1 Environmental Impact Assessment processes
Source: Based on European Union (2001); DUAP (1995); Thomas (1998).

The EIA process is often more complex than shown in the figure with feedback loops and sidetracks especially when development is controversial and the public and politicians become heavily involved.

The centre process stream illustrates the main sequence. Early and continuing consultation with the relevant environmental and planning authorities and the affected stakeholders (Chapter 19) is important for a well managed EIA, one in which the organisation remains in control. The author has added two aspects that should be linked to EIA: environmental design management (Chapter 23) and the organisation's EMS (Chapter 12).

Figure 25.1 shows a separate stream for environmental reviews (Thomas, 1998). These may be required by local as distinct from central authorities or by the requirements of an organisation's EMS.

Scope determination

Developments and projects arise from governmental planning (a new health services supply policy, a new harbour, a new railway or private planning, a new product, a new factory or a new urban land development). In the process of establishing the need or feasibility of the development, an organisation's internal planning generates schemes which will ultimately have an impact on the environment. As scope is refined alternatives are raised and assessed. Environmental issues are important considerations even at this early stage.

In defining the development, an organisation should (as well as assessing project management issues of scope, cost, time and resources for the project), look at the environmental issues. These include: relevant legislation; the statutory approvals required; the formal processes necessary to gain these approvals; the organisations and people (private and governmental) to consult; and the resources and the timing of the EIA process.

Screening for EIS

The proponent needs to make a decision whether an EIS is necessary and this is done in conjunction with the relevant approval authorities and by understanding the planning and environmental legislation of a country. Figure 25.1 shows this process and the decision point that leads to either an EIS or an environmental review process.

25.4   EIS scope

The scope of the EIS is determined by the scope of the development or project, the extent of the approvals necessary and the extent and severity of the environmental impacts.

The EIS framework is driven by the relevant legislation and regulations (e.g. planning, pollution, environmental), the planning instruments (country or state environmental planning policies, regional environmental plans, local environmental plans), the statutory authorities' requirements (consent authorities, other approval authorities), the approval processes for development approvals (applications, certificates, permits, licences, approvals, environmental management plans) and public and stakeholders participation.

The requirements of the relevant legislation need to be addressed in the EIS and could include: urban and rural planning, flora and fauna, national parks and wildlife, threatened species and communities, contaminated land management, rivers, foreshores, heritage, pollution, waste generation, economic issues, social issues and environmental resources issues.

25.5 Planning and managing an EIS project

An organisation needs to carefully plan the EIS preparation and if in-house resources are insufficient will need to use consultants e.g. to manage the whole project or to provide specialist studies (see Figure 25.2).

An EIS project lends itself easily to the use of the tools of project management. These include: management of the scope of the EIS project; control of variations to scope; control of time – planning, scheduling, progress monitoring; control of cost – budgeting, commitments and expenditure; identification and management of the resources necessary; identification and managing of the stakeholders inputs; managing the client organisation or organisational group input; identification and liaison with relevant statutory authorities;establishing the meeting schedule, control of meetings, control of documentation flow and content; quality control of documentation and technical content; and monitoring and management of approval process after the application is submitted.

25.6 Specialist studies

The specialist studies whether undertaken by the organisation or its EIS consultant would typically need to cover the following topics:

- legislation;
- land use;
- economics of the project at local, regional and national scales;
- employment, traffic and transport both during construction and operation;
- social including people in the area and remote;
- stakeholders including the organisation's own personnel (managers and board), suppliers, customers, shareholders, community, special interest groups, public authorities;
- community concerns including health of people in the development and in its vicinity;
- pollution existing or likely of air, water, land (existing or potential),

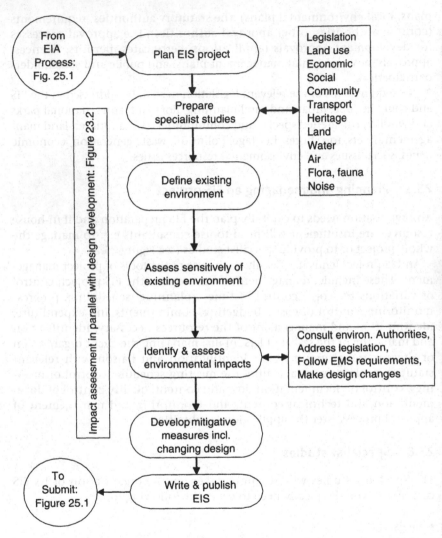

Figure 25.2   EIS preparation
Source: Based on European Union (2001); DUAP (1995); Thomas (1998); Glasson et al (1999).

- heritage;
- land characteristics e.g. soils, geology, ground water and land rehabilitation and landscaping;
- water;
- air;
- flora and fauna including threatened species, threatened communities;
- noise;

- waste management and minimisation; and
- the opportunities for application of the principles of ecological sustainability.

These studies help define the existing environment and together with the scientific and engineering analysis of the project (especially the material and energy use and the waste and pollution outputs) allow the impacts of the project on the environment (natural, heritage and social) to be predicted and mitigative measures to be planned.

Impact prediction and assessment

Impact prediction and assessment is one of the most important parts of an EIS and needs to be based on good scientific and engineering information. If the data is not robust the EIS should adopt the precautionary principle and qualify the data, the prediction and the assessment especially if value judgements are involved.

Mitigative measures and project design

During the process of developing and writing the EIS, mitigative measures are identified by the organisation, in response to requirements of the approval authorities and community or stakeholder concerns (Chapter 19). For an organisation with an EMS and a commitment to sustainability it is better if the major aspects of mitigation are developed by the organisation during the early stages of realisation of the development or project. This can result in reduced impacts.

25.7 Submission documentation

The type and detail of the submission documentation depends on significance and size of the impact, applicable legislation, requirements of the approval authority, the types of impact statements (e.g. environmental impact statement or environmental review), the need for environmental management plans, EIA monitoring and reporting.

Typically it would include: description of the development or project; description of the existing environment both natural and human; identification of the requirements of the relevant legislation and special requirements of approval authorities; the environmental impact of the project; justification of project including alternatives considered; community or stakeholder concerns; and mitigative measures proposed by the proponent.

25.8 Exhibition and approval process

For effective management of the EIA process by the organisation it is important to obtain stakeholder involvement as early as possible (Chapter 19) including involvement from the organisation's own personnel (managers and Board), suppliers, customers, shareholders, effected community, special interest groups, public authorities and approval authorities. Early meetings with statutory authorities are required to establish relationships, identify legal requirements and identify specific requirements.

Engagement of the key stakeholders at the start will help define and agree processes, enable a preliminary assessment to be outlined, enable the assessment to be integrated with design so constraints can be identified, design revised, impacts lessened and to identify and justify the benefits for government, community and the organisation.

25.9 Conclusion

Since the early 1970s many countries have adopted EIA procedures and there are many ways of undertaking EIA (Thomas, 1998). This section has outlined a simplified form of the EIA process as it applies to an organisation undertaking a development or project. The organisation could be public or private, large or small. It focuses on the management of the process rather than the technical detail of how to research and write an EIS. This can be found in the references and the books listed under further reading. The detailed requirements for the EIS are often outlined by the relevant legislation and in guidelines prepared by the approval authorities.

25.10 Further reading

Morris and Therivel (2001); Singleton et al (1999); Wood (1995); Thomas and Elliott (2005).

25.11 Questions

- Does EIA contribute to the preservation of the natural environment?
- To what extent does EIA document and justify decisions already made by an organisation?
- Should the development proponent prepare its own environmental impact statement or should an independent third party prepare it?
- Identify a controversial EIA in your country and assess whether the community have been properly consulted and changes made in response to their concerns.

26
Environmental Auditing
Robert Staib

26.1 Introduction

This chapter outlines the processes involved in environmental auditing from an organisational point of view making the distinction between the internal and external auditing and between system and performance audits as shown in Figure 26.1.

The internal audit is an important tool for the control and management of the organisation while the external audit can be used for the verification

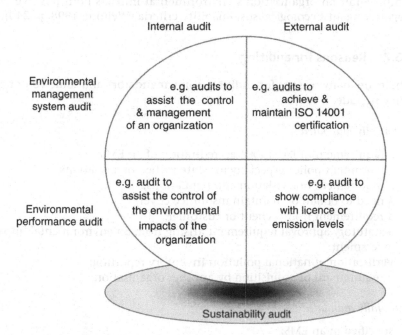

Figure 26.1 Environmental audit categories

by external stakeholders of an organisation's environmental compliance with legislation, with community, customer or client expectations and with its ISO 14001 certification (Section 26.3). The internal audit can be undertaken by internal organisational resources though external resources are often used for objectivity or to complement internal resources. The external audit is undertaken by an independent external party.

The distinction between an audit of the organisation's environmental management system and management processes (e.g. its EMS) and an audit of its environmental performance or environmental impact (Figure 26.1) is important to understand (Hillary, 1998). The first evaluates an organisational system, the second the physical environment and both aspects should be included in an organisation's audit program (Sullivan and Wyndham, 2001).

The International Organization for Standardization's definition of environmental auditing is the 'systematic, documented verification process of objectively obtaining and *evaluating* audit evidence to determine whether specified environmental activities, events, conditions, management systems, or information about these matters conform with audit criteria, and communicating the results of this process to the client'. The key word is *evaluating* (ISO, 2002).

Another category of audit that organisations should consider is a sustainability audit to assess an organisation's progress towards sustainability. This would audit an organisation's environmental impacts both positive and negative against ecological sustainability criteria (Welford, 1998, p. 240).

26.2 Reasons for auditing

There are many reasons for auditing, some mandatory and some voluntary. They include:

Externally mandated

- a requirement of ISO 14001 accreditation of an EMS;
- government policy especially for state owned organisations;
- a requirement of legislation and regulations;
- a need to obtain or maintain insurance cover;
- a requirement from a client or customer;
- a statutory approval requirement arising from an environmental impact assessment;
- verification of national pollution inventory reporting;
- a requirement of acquisition by another organisation;

Internally mandated

- specified in an EMS;
- a requirement of a supply or construction contract;

Voluntary

- confirmation of compliance with legislation and regulations;
- environmental risk assessment;
- evaluation of environmental impacts;
- evaluation of environmental performance;
- management audits for due diligence;
- assessment of future environmental options;
- support for green product or organisational marketing claims;
- system integration e.g. OHS, environmental, quality, risk management; and
- measurement of progress towards an ecologically sustainable organisation.

(Staib, 1993–2004, 1997–2000; Environmental Protection Authority, 1996; Welford, 1998; Barrow, 1999).

Some legislation and government regulations will have mandatory auditing requirements especially for government owned organisations. Some of the items under *'Externally mandated'* may have a legislated basis. EIA approvals and operating licences issued by statutory authorities may have mandatory auditing requirements with also a need for an auditor independent of the organisation. An organisation needs to identify all legal auditing requirements to establish a base level of auditing.

26.3   Standard codes

Environmental auditing probably originated in the United States of America in the 1970s but has been accepted and developed in many countries since then. Early directions were provided by the International Chamber of Commerce (Hillary, 1998; Barrow, 1999) and the USA EPA's Environmental Auditing Policy Statement in 1986 (Eccleston, 1999). Standard approaches have been developed initially through British Standard BS 7750, through the Eco-Management and Audit Scheme, the international standard ISO 14010 (ISO, 1996; Barrow, 1999, p. 70) and latterly ISO 19011: 2002 (ISO, 2002).

The ISO environmental standards are now widely accepted and applied. They are standards that describe the processes to be followed and environmental aspects to consider but do not proscribe levels of impact or quantitative environmental goals to achieve. These are left to the organisation and its stakeholders to develop. The ISO standard for auditing (now in one volume after being initially in three) is 'Guidelines for quality and/or environmental management systems auditing ISO 19011: 2002'. An associated code is ISO 14031 which covers Environmental Performance Evaluation (ISO, 1999a).

Some countries have adopted these as written and others have incorporated them into their own standards system e.g. Australia and New Zealand through the AS/NZS ISO series.

26.4   Aims

An audit is only a sampling of the organisation's environmental management and achievements even though an audit may be comprehensive, extend over days and review years of recorded data and information. Therefore it is important for the aims and scope of each audit to be clearly defined.

The reasons for audits (Section 26.2) will dictate the aims or objectives that an organisation needs to set for its audits and will also help it establish the scope and the timing of the organisation's audit program. Aims can include:

- establishing a base of environmental information preparatory to establishing an EMS;
- confirming compliance with and effectiveness of the organisation's EMS;
- assessing and verifying compliance with standards, legislation, and organisational policies and objectives;
- identifying problems and corrective actions;
- helping an organisation to formulate and change environmental policy, environmental strategy and environmental programs;
- measuring the environmental impact and ongoing environmental performance of development projects and operations; and
- Providing a base of information for communication, reporting and decision making.
(Staib, 1993–2004, 1997–2000; Welford, 1998, Chapter 7).

26.5   Principles for effective audits

It is more important to identify why an audit is required (Section 26.2), to be clear about the aims of the audit (Section 26.4) and to follow good auditing principles (Section 26.5) than to worry unduly about the correct name or type of an audit.

There will always be the need for ad-hoc audits but organisations will need to establish an ongoing audit program with regular audits to show continuing compliance, continual improvement and to provide current information for the management of environmental aspects within the organisation and in its interface with external stakeholders.

Some of the important principles for effective audits are:

- commitment of top management to the audit process, the audit program and to the implementation of the audit recommendations;
- statement of clear aims and objectives and evaluation criteria for each audit;
- detailed planning of all aspects of the audit process and its activities;

- establishment of an audit team that is proficient, expert in the relevant areas, can act independent and with confidentiality;
- involvement of the auditees and auditee organisation at an early date;
- undertaking the audit in a professional manner that is not disruptive to the organisation's operation and following safe working practices while on site;
- review of both the management systems and the environmental performance (Refer Figure 26.1);
- gathering of sufficient creditable evidence to draw valid conclusions;
- production of a well constructed report that is clear and concise and makes practicable recommendations and includes an executive summary listing each recommendation clearly; and
- identification to the auditee, at the conclusion of the audit, of any non-conformance that could lead to prosecution.

(Staib, 1993–2004, 1997–2000; Sullivan and Wyndham, 2001; Sweet et al, 2003).

26.6 Implementing an audit

The activities involved in the implementation of an environmental audit will depend on the scope, size and aims of audit as well as the frequency of auditing. Time spent in planning the audit is time well spent and will ensure that the appropriate aims and objectives are set and a competent and qualified audit team established. Figure 26.2 shows some of the main activities necessary.

The audit team needs to gather sufficient verified information (through interviews, data collection and site visits) to enable it to arrive at robust recommendations that will be accepted by the organisation and its stakeholders. The on-site activities cover both the audit of the environmental management system and the environmental performance. The audit team also needs to evaluate the adequacy of the information collected to determine whether they need to seek further tests or monitoring to compliment that collected during the audit.

The opening and closing meetings enable the audit team leader to establish communications with the audit team and the audited personnel. During the closing meeting and after reading the audit report, the audited organisation needs to be convinced by the audit team that it should implement the recommendations to achieve compliance and better environmental performance (Staib, 1993–2004).

Figure 26.2 shows the audit process starting from within an existing audit program though it would be similar for an ad-hoc one-off or initial audit. The audit ends with the audit report, which implies that the audited organisation has the responsibility for establishing the action plan and for implementing this plan. The auditor can be involved in providing advice and reauditing to confirm successful implementation.

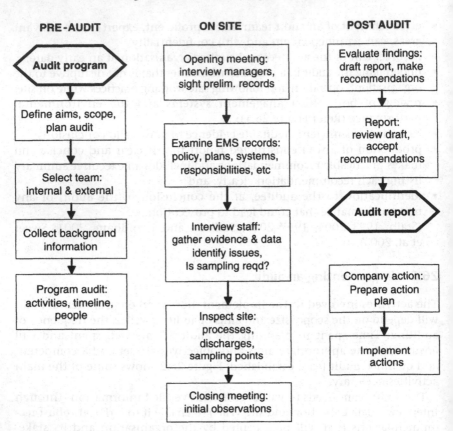

Figure 26.2 Environmental audit activities
Source: Based on Environmental Protection Authority (1996, p. 38); Welford (1998, p. 127); Staib (1993–2004, 1997–2000).

26.7 Audit program and systems integration

As auditing is a part of an organisation's environmental management system, it will be necessary to establish an audit program that schedules regular audits on all aspects of the organisation's interface with the environment and external stakeholders.

This would include many different types of audits that evaluate the organisation's EMS, legislative compliance, compliance with statutory approvals and licences, achievement of environmental goals, maintain contact with the changing needs and demands of the market place and stakeholders.

Increasingly (Hortensius et al, 2004; Mattieu, 2003), organisations are looking to establish integrated management systems pulling together the strands of their separate systems e.g. environmental management, occupational health and safety management, quality management and other management systems. This will result in a need to rationalise auditing and to integrate some of the common aspects in combined audits.

26.8   Auditors

An organisation needs to address several issues when it comes to establishing an audit team. Larger organisations may have resources in-house to staff an audit team with the appropriate level of skills but if the audit report is to be provided to external stakeholders it is better to include independent people on the audit team to establish an independent creditability. Even for internal audits if an organisation wants to establish credibility or take a fresh look at its operations it could use the services of an external auditor or audit organisation.

The ISO standard (ISO, 2002) defines a process for the accreditation of auditors. If the audit is totally external to the organisation e.g. externally mandated, the audit team will be independent of the organisation being audited and is more likely to utilise accredited auditors.

Auditors of an EMS may be generalists with a management or technical background whilst an environmental performance audit would need technical specialists in pollution measurement, industrial processes, waste management or legal training.

26.9   Conclusion

Environmental auditing has been growing since the early 1970s with both major companies and small and medium companies undertaking environmental audits and being audited. This chapter has outlined some of the reasons for auditing. It provides a framework for planning and managing an audit and highlights the need to have a program of regular audits. Audits can be intrusive but properly managed they will assist an organisation to avoid prosecution, identify its environmental performance, improve its environmental management and ultimately reduce its impact on the environment.

26.10   Further reading

Brown (1993); ISO (2004a).

Case study: Delta Electricity

Delta Electricity, a public owned company, is a major producer of electricity for the state of New South Wales, Australia with 4,700 MW of installed capacity of which about 90% is from coal firing and the remainder from hydroelectricity. It produced its first environmental report in 1999. It operates under a pollution licence issued by the New South Wales Environment Protection Authority that regulates the discharge of pollutants to air, water and land. It is also signatory to a number of agreements that require it to verify its environmental performance through audit: the Electricity Supply Association of Australia, the voluntary Greenhouse Challenge Agreement with the Australian federal government, the federal government's National Pollutant Inventory as well as its ownership by the state of New South Wales.

Delta Electricity has engaged the services of external consultants to undertake its main audits, which cover environmental management, facilities and process, compliance, due diligence and its EMS. Findings help support its reporting requirements, identify environmental risks and maintain its ISO 14001 accreditation. An earlier external audit assisted Delta Electricity to identify the main environmental aspects and impacts of its power station operations, to develop an in-house environmental training package and to implement a computerised EMS (Delta Electricity, 2000, 2003).

26.11   Questions

- As public and private organisations use resources and create waste and pollution, should they be required to undertake annual independent audits of their environmental management systems and environmental performance and the results made public?
- Should organisations that make green marketing claims be subject to an environmental audit by the relevant consumer protection authority of a country to verify their claims?
- Choose a significant organisation that has many environmental issues and prepare a program of audits with aims and objectives and identify the auditor skills required in the audit teams.

References

C. A. Adams, 'The nature and processes of corporate reporting on ethical issues', *The Chartered Institute of Management Accountants*, 4 (1999) 1–64.

C. A. Adams, 'Internal organisational factors influencing corporate social and ethical reporting: Beyond current theorising', *Accounting, Auditing and Accountability Journal*, 15 (2) (2002) 223–250.

C. A. Adams and G. F. Harte, 'The changing portrayal of the employment of women in British banks' and retail companies' corporate annual reports', *Accounting, Organizations and Society*, 23 (8) (1998) 781–812.

N. Adler and M. Tushman, *Competing by Design: the Power of Organisational Architecture*, (Oxford: Oxford University Press, 1997), p. 240.

P. Adler and S. Kwon, 'Social Capital: Prospects for a new concept', *Academy of Management Review*, 27 (1) (2002) 17–40.

I. Ajzen, 'On Behaving in Accordance with One's Attitudes', in H. E. T. Zanna M. P., Herman C. P., *Consistency in Social Behaviour*, (Hillsdale, New Jersey: Lawrence Erlbaum Associates, 1982), pp. 3–15.

I. Ajzen, 'Nature and Operation of Attitudes', *Annual Review of Psychology*, 52 (2001) 27–58.

I. Ajzen and M. Fishbein, 'Attitude – Behaviour Relations, A Theoretical Analysis and Review of Empirical Research', *Psychological Bulletin*, 84 (1977) 888–918.

I. Ajzen and M. Fishbein, *Understanding Attitudes and Predicting Social Behaviour*, (Englewood-Cliffs, New Jersey: Prentice-Hall, 1980), p. 278.

L. R. Alm, *Crossing Borders, Crossing Boundaries: The Role of Scientists in the U.S. Acid Rain Debate*, (London: Praeger Publishers, 2000), p. 147.

V. Anand, 'Building blocks of corporate reputation – Social responsibility initiatives', *Corporate Reputation Review*, 5 (1) (2002) 71–74.

M. Anderson and I. Massa, 'Ecological Modernization – Origins, Dilemmas and Future Directions', *Journal of Environmental Policy and Planning*, 2 (2000) 337–345.

Anonymous, 'Green Cleaners', *CHOICE*, (1990 September) 10–14.

Anonymous, 'Spurts and Starts: Corporate Role in '90s Environmentalism Hardly Consistent', *Advertising Age*, 62 (46) (1991) GR14–GR16.

Aquatec Environmental Consultants, *Economic Assessment of Environmental Impacts, Annexure to the NSW Government Guidelines for Economic Appraisal (unpublished report)*, (Sydney, Australia: NSW Government (unpublished), 1996), p. 76.

Aristotle, *Politics (with an English translation by H. Rackham)*, (London: Heinemann, 1967), p. 683.

S. R. Arnstein, 'A ladder of citizen participation', *Journal of the American Institute of Planners*, 35 (1969) 216–224.

S. Ashley, 'Designing for the Environment', *Mechanical Engineering*, 115 (3) (1993) 52–55.

Asian Development Bank, *Guidelines: Policy Integration and Strategic and Action Planning for the Achievement of Cleaner Production*, (Manila, Philippines: Asian Development Bank, 2002), http://www.uneptie.org/pc/cp/reportspdf/ADBCPPolicy.pdf, viewed 27 July 2004.

Association of British Insurers, *Disclosure Guidelines for Social Responsibility*, (London: ABI, 2002), p. 6.

K. Auguston, 'In search of pallet solutions', *Modern Materials Handling*, 51 (9) (1996) 38–41.

Australasian Reporting Awards (ARA), *Australasian Reporting Awards*, (Sydney: ARA, 2004), http://www.arawards.com.au/, viewed 19 June 2004.

Australian Industry Group, *About Australian Industry Group, Background and history*, (Canberra, Australia: Australian Industry Group, 2004), http://www.aigroup.asn.au, viewed 22 May 2004.

Australian Securities & Investments Commission (ASIC), *Section 1013DA disclosure guidelines*, (Canberra: ASIC, 2003), p. 22.

R. Ayres, G. Ferrer and T. van Leynseele, 'Eco-Efficency, Asset Recovery and Remanufacturing', *European Management Journal*, 15 (5) (1997) 557–574.

R. Bailey, *Earth Report 2000, Revisiting the True State of the Planet*, (New York: McGraw-Hill, 2000), p. 362.

J.-P. Barde and N. A. Braathen, *Environmentally Related Levies. Paper prepared for the Conference on Excise Taxation, 11–12 April 2002*, (The Hague, The Netherlands: Rotterdam School of Economics, 2002), http://www.few.eur.nl/few/research/ocfeb/excisetaxpolicy/papers.htm, viewed 25 October 2004.

J. Barrett and A. Scott, 'The Ecological Footprint: A Metric for Corporate Sustainability', *Corporate Environmental Strategy*, 8 (4) (2001) 316–325.

C. J. Barrow, *Environmental Management Principles and Practice*, (London: Routledge, 1999), p. 326.

G. Bates, *Environmental Law in Australia, 4th edn*, (Australia: Butterworths, 1995), p. 534.

M. Bazerman and A. Hoffman, 'Sources of Environmentally Destructive Behavior: Individual, Organizational, and Institutional Perspectives', *Research in Organizational Behavior*, 21 (1999) 39–79.

R. Beale, 'Turf protection: Conflict between authorities', *Australian Journal of Public Administration*, 54 (2) (1995) 143–148.

U. Beck, *World Risk Society*, (Cambridge: Polity Press, 1999), p. 184.

S. Beder, *Global Spin: The Corporate Assault on Environmentalism*, (Carlton, Victoria: Scribe Books, 1997), p. 288.

T. Beer, *Preliminary Life Cycle Assessment Road Construction materials Homebush Bay (CSIRO Ref SB/1/292F2)*, (Sydney: Commonwealth Scientific and Industrial Research Organisation – unpublished, 1996), p. 24.

S. Behrendt, C. Jasch, M. C. Peneda and H. van Weenen, *Life Cycle Design A Manual for Small and Medium-Sized Enterprises*, (Berlin: Springer-Verlag, 1997), p. 190.

M. Belz, 'Eco-Marketing 2005', in M. Charter and M. J. Polonsky, *Greener Marketing: A Global Perspective to Greening Marketing Practice Second Edition*, (Sheffield, UK: Greenleaf Publishing, 1999), pp. 84–94.

M. Bengisson and B. Steen, 'Weighting in LCA – approaches and applications', *Environmental Progress*, 19 (2) (2000) 101–109.

S. Benn and D. Dunphy, 'A case study in corporate sustainability: Fuji XEROX Eco Manufacturing Centre', *Innovation Management Policy and Practice*, 6 (2) (2004a) 258–268.

S. Benn, D. Dunphy and P. Crittenden, *Carlton United Brewery: Kent Street Ultimo, A Case Study* (Sydney: UTS, unpublished, 2004b).

F. Berkes and C. Folke, 'Investing in cultural capital for sustainable use of natural capital', in A. Jannson, M. Hammer, C. Folke and R. Constanza, *Investing in natural Capital – the ecological approach to sustainability*, (Washington, DC: Island Press, 1994), p. 128–150.

S. Bernhut and P. Bansal, 'Corporate social responsibility', *IVEY Business Journal*, 66 (4) (2002) 18–19, 59.

V. N. Bhat, 'Green Marketing Begins with Green Design', *Journal of Business and Industrial Marketing*, 8 (4) (1993) 26–31.

J. Birkeland, *Design for Sustainability: A Sourcebook of Integrated Eco-logical Solutions*, (London: Earthscan, 2002), p. 288.

T. Blair, *Sustainable Development Helps the Poorest (speech delivered on 24 February 2003)*, (London: UK Labour Party, 2003), http://www.labour.org.uk/news/tbsd, viewed 27 September 2004.

J. Bowman and C. Davis, 'Industry and the Environment: Chief Executive Officer Attitudes, 1976 and 1986', *Environmental Management*, 13 (2) (1989) 243–249.

J. W. Broderick, D. R. Lavoie and A. J. Perel, 'Environmental risk management and the role of environmental insurance', *Environmental Quality Management*, 10 (1) (2000) 3–12.

G. Brown, 'Now you can invest without sacrificing profit for principle', *Asset Financial Management*, May 1 (2003).

G. A. Brown, *Environmental Audit Guidebook*, (Melbourne: Centre for Professional Development, 1993), p. various.

J. Brown and P. Duguid, 'Organisational learning and communities of practice: Toward a unified view of working, learning and innovation', *Organisation Science*, 2 (1) (1991) 40–57.

G. H. Brundtland, *Our Common Future: World commission on environment and development*, (Oxford: Oxford University Press, 1987), p. 400.

A. Bryden, ' Viewpoint of International Standards Organisation's New Secretary-General,' *Management Systems The International Review of ISO 9000 and ISO 14000*, 3 (3) (2003) 15–17.

D. Buchan, 'A bit of a surprise: BP: The oil giant's environmental efforts have won plaudits from an unexpected source', *Financial Times*, Dec 17 (2001) 8.

J. Buchanan and P. Ryan, 'Private and Public Faces of Ethics: Convergence and Divergence', *International Institute for Public Ethics Conference, Reconstructing 'the Public Interest' in a Globalising World: Business, the Professions and the Public Sector*, (Brisbane: International Institute for Public Ethics, 2002) 1–30. http://www.iipe.org/conference2002/papers/Buchanan.pdf, viewed 10 April 2005.

N. Buck, 'Integrating compliance management systems', *Keeping Good Companies*, 53 (7) (2001) 410–411.

BuildingGreen Inc., 'Life Cycle Assessment for Buildings – Seeking the Holy Grail', *Environmental Building News*, 11 (3) (2002).

Bureau of International Recycling, *Tires Recycling*, (Brussels: BIR, 2004), http://www.bir.org/aboutrecycling/tyres.asp, viewed 1 July 2004.

R. L. Burritt and L. S. Cummings, 'Accounting for Biological Assets – the Experience of an Australian Conservation Company', *Asian Review of Accounting*, 10 (2) (2002) 17–42.

R. Butler, 'The end of the pipe: reducing emissions is a sustainable goal, however, in focusing on immediate technologies is long-term impact research being overlooked', *Chemistry and Industry*, August 5 (2002) 14–15.

M. Byrne and M. J. Polonsky, 'Impediments to Consumer Adoption of Sustainable Transportation: Alternative Fuel Vehicles', *International Journal of Production and Operation Management*, 21 (12) (2001) 1521–1538.

E. Cagno, P. Trucco and L. Tardini, 'Cleaner production and profitability: analysis of 134 industrial pollution prevention (P2) project reports', *Journal of Cleaner Production* in press (*on-line* http://www.sciencedirect.com), 13 (6) (2005) 1–13.

J. B. Callicott, 'Non-anthropocentric Value Theory and Environmental Ethics', *American Philosophical Quarterly*, 2114 (1984) 299–308.

J. B. Callicott, *Earth's Insights: a multicultural survey of ecological ethics from the Mediterranean Basin to the Australian outback*, (California: University of California Press, 1994), p. 292.

J. B. Callicott, *Beyond the Land Ethic: More essays in environmental philosophy*, (Albany, New York: State University of New York Press, 1999), p. 427.

S. Campbell and S. Green, 'Net Alters Face of Distribution', *Computer Reseller News*, 869 November 15 (1999) 97.

L. Carlson, S. Grove, N. Kangun and M. J. Polonsky, 'An International Comparison of Environmental Advertising: Substantive vs Associative Claims', *Journal of Macromarketing*, 16 (2) (1996) 57–68.

L. Carson and K. Gelber, *Ideas for Community Consultation: A discussion on principles and procedures for making consultation work*, (Sydney: NSW Department of Urban Affairs and Planning, 2001), http://www.iplan.nsw.gov.au/engagement/stories/docs/ideasforconsult.pdf, viewed 28 December 2004.

R. L. Carson, *Silent Spring*, (London: H. Hamilton, 1963), p. 304.

P. Cebon, 'Corporate Obstacles to Pollution Prevention', *EPA Journal*, (1993) 20–22.

R. A. Chechile and S. Carlisle, *Environmental Decision Making: A Multidisciplinary Perspective*, (New York: Van Nostrum Reinhold, 1991), p. 296.

P. Christoff, 'Ecological Modernisation, Ecological Modernity', *Environmental Politics*, 5 (1996) 476–500.

B. W. Clapp, *An Environmental History of Britain since the Industrial Revolution*, (London: Longman, 1994), p. 235.

T. Clarke, *Theories of Governance*, (London: Routledge, 2004), p. 384.

R. T. Clemen and T. Reilly, *Making Hard Decisions with Decision Tools*, (Pacific Grove, California: Duxbury Press, 2001), p. 733.

R. Clift, 'Overview Clean Technology – The Idea and the Practice', *Journal of Chemical Technology & Biotechnology*, 68 (4) (1997) 347–350.

M. Cohen, 'Environmental Crime and Punishment: Legal/Economic Theory and Empirical Evidence on Enforcement of Federal Environmental Statutes', *Journal of Criminal Law and Criminology*, 82 (1992) 1054–1108.

E. Cohen-Rosenthal, 'A walk on the human side of industrial ecology', *American Behavioral Scientist*, 44 (2) (2000) 245–264.

D. Coles and D. D. Green, *Do UK Pension Funds Invest Responsibly? A Survey of Current Practice on Socially Responsible Investment*, (London: Just Pensions, 2002), p. 16.

J. Connolly, P. McDonagh, M. Polonsky and A. Prothero, 'Green Marketing and Green Consumers: Exploring the Myths', in D. Annandale, Marinova and J. Phillimore, *International Handbook on Environmental Technology Management (Forthcoming)*, (Cheltenham, UK: Edward Elgar, 2004).

F. Consoli, D. Allen, I. Boustead, J. Fava, W. Franklin, A. A. Jensen, N. de Oude, R. Parrish, R. Perriman, D. Postlethwaite, B. Quay, J. Seguin and B. Vigon, *Guidelines for Life-Cycle Assessment: A Code of Practice*, (Brussels and Pensacola: Society for Environmental Toxicology and Chemistry, 1993), p. 73.

B. J. Cook, 'The politics of market-based environmental regulation: continuity and change in air pollution control policy conflict', *Social Science Quarterly*, 83 (2002) 156–167.

Coopers & Lybrand, *Environmental Management Practices: A survey of major Australian organisations*, (Sydney/Melbourne: Coopers & Lybrand Consultants, 1991).

D. Cottam, 'A Green Policy Committed to Print: Kyocera's Cartridge-Free Laser Printer', *Greener Management International*, 5 (1994) 61–66.

J. Court, C. Wright and A. Guthrie, 'Environmental Assessment and Sustainability: Are We Ready for the Challenge', *Australian Journal of Environmental Management*, 3 (1) (1996) 42–57.

D. Coward, *Out of Sight, Sydney's Environmental History 1851–1981*, (Canberra: Department of Economic History, Australian National University, 1988), p. 328.

S. J. Cowell, R. Fairman and R. E. Lofstedt, 'Use of Risk Assessment and Life Cycle Assessment in Decision Making: A Common Policy Research Agenda', *Risk Analysis: An International Journal*, 22 (5) (2002) 879–895.

A. Crain, 'Facing the Backlash: Green Marketing and Strategic Reorientation in the 1990's', *Journal of Strategic Marketing*, 8 (2000) 277–296.

A. Crane, 'Exploring Green Alliances', *Journal of Marketing Management*, 14 (4) (1998) 559–580.

D. Craney, *The Environment Movement in NSW 1969 to 1978*, Master of Environmental Studies, (Ryde, Sydney: Macquarie University, 1980), p. 494.

R. Cranston, 'Regulation and Deregulation: General Issues', in R. Tomasic, *Business Regulation in Australia*, (Sydney: CCH Australia, 1984), pp. 33–41.

CSIRO Urban Water (CSIRO) and Institute of Sustainable Futures, *Edmonston Park Feasibility Report (unpublished)*, (Sydney: CSIRO Australia, University of Technology Sydney, 2002), p. 171.

M. A. Curran and S. Young, 'Report from the EPA Conference on Streamlining LCA', *International Journal of Life Cycle Assessment*, 1 (1) (1996) 57–60.

K. Cussen, 'Aesthetics and environmental argument', *Essays in Philosophy*, 3 (1) (2002) http://www.humboldt.edu/~essays/cussen.html, viewed 29 November 2004.

B. Daily and S. Huang, 'Achieving sustainability through attention to human resource factors in environmental management', *Journal of Operations and Production Management*, 21 (12) (2001) 1539–1552.

H. D. Dale and M. R. English, *Tools to Aid Environmental Decision Making*, (New York: Springer-Verlag, 1998), p. 342.

R. Dalton and M. Kuechler, *Challenging the Political Order: New Social and Political Movements in Western Democracies*, (Cambridge: Polity Press, 1990), p. 329.

G. Davis, J. Wanna, J. Warhurst and P. Weller, *Public Policy in Australia*, (Sydney: Allen and Unwin, 1993a), p. 294.

J. Davis, 'Strategies for Environmental Advertising', *Journal of Consumer Marketing*, 10 (2) (1993b) 19–36.

A. de Geus, 'The Living Company', *Harvard Business Review*, 75 (1997) 51–59.

Delta Electricity, *Environmental Review 1999–2000*, (Sydney: Delta Electricity, 2000), p. 20.

Delta Electricity, *2002/2003 Environmental Report in the Annual Report*, (Sydney: Delta Electricity, 2003), http://www.de.com.au/ArticleDocuments/Delta_AR03%209-16.pdf, viewed 15 April 2004.

Deni Greene Consulting Services, *Socially Responsible Investment in Australia – 2002. Benchmarking Survey Conducted for the Ethical Investment Association*, (Sydney: Ethical Investment Association, 2002), p. 30.

B. Dennis and C. P. Neck, 'Body Shop International: an exploration of corporate social responsibility', *Management Decision*, 36 (9/10) (1998) 649–654.

M. Dickson, 'Of ethics, indices and the wisdom of the easter bunny: A flurry of events over the past few days demonstrate that companies will have to pay increasing attention to their social responsibilities, whether they like it or not', *Financial Times*, April 14 (2001) 11.

C. Dimitri and C. Greene, *Organic Food Industry Taps Growing American Market*, (Washington D. C.: United States Department of Agriculture, 2002), www.ers.usda.gov/briefing/, viewed 1 August 2004.

R. Dinham, *Personal communication with architect for Olympic Facilities*, (Sydney: 1998).

E. Dohlman, 'The trade effects of environmental regulation', *OECD Observer*, 162 (1990) 28–32.

B. Doppelt, *Leading Change toward Sustainability*, (Sheffield: Greenleaf Publishing, 2003), p. 272.

A. Downs, 'Up and down with ecology? the "issue attention cycle"', *Public Interest*, 28 (1972) 38–59.

D. Doyle and D. McEachern, *Environment and Politics*, (London: Routledge, 1998), p. 206.

T. Doyle and A. Kellow, *Environmental Politics and Policy Making in Australia*, (Melbourne: Macmillan, 1995), p. 296.

M. E. Drumwright, 'Socially Responsible Organizational Buying: Environmental Concern as a Noneconomic Buying Criterion', *Journal of Marketing*, 58 (3) (1994) 1–19.

J. T. Dryzek, *The Politics of the Earth: Environmental Discourses*, (Oxford: Oxford University Press, 1997), p. 220.

DUAP, *Is an EIS Required? Best Practice Guidelines for Part 5 of the Environmental Planning and Assessment Act 1979*, (Sydney, NSW: Department of Urban Affairs and Planning, 1995), p. 59.

R. E. Dunlap, 'Trends in Public Opinion Towards Environmental Issues: 1965–1990', *Society and Natural Resources*, 4 (1991a) 285–312.

R. E. Dunlap, 'Public Opinion in the 1980s Clear Consensus, Ambiguous Commitment', *Environment*, 33 (8) (1991b) 10–15.

D. Dunphy, A. Griffiths and S. Benn, *Organisational Change for Corporate Sustainability*, (London: Routledge, 2003), p. 315.

Earth Sanctuaries Ltd, *Annual Reports, 1995–2001*, (Stirling, South Australia: Earth Sanctuaries, 2004), http://www.esl.com.au/, viewed 19 June 2004.

R. J. Eaton, 'Free enterprise in the 21st century: how much government interference is necessary', *Vital Speeches of the Day*, 61 (16) (1995) 509–513.

C. H. Eccleston, *The NEPA Planning Process*, (New York: John Wiley & Sons, 1999), p. 396.

Eco-labels, *Labels, Eco-Labels*, (New York: Consumers Union, 2004), www.eco-labels.org, viewed 1 July 2004.

R. Edwards, 'How to take corporate responsibility: Just do it', *The Sunday Herald*, Aug 18 (2002) 4.

J. R. Ehrenfeld, 'Industrial ecology: paradigm shift or normal science?' *American Behavioural Scientist*, 44 (2) (2000) 229–244.

P. R. Ehrlich, A. H. Ehrlich and J. P. Holdren, *Ecoscience Population Resources Environment*, (San Francisco: W. H. Freeman Company, 1977), p. 1051.

J. Elkington, *Cannibals with forks: The triple bottom line of 21st Century business*, (Gabriola Island: New Society Publishers, 1998), p. 407.

J. Elkington, *The Chrysalis Economy*, (Oxford: Capstone, 2001), p. 284.

L. Elliott, *The Global Politics of the Environment*, (London: Macmillan, 1998), p. 331.

Environment Australia, *Public environmental reporting: An Australian approach*, (Sydney: Australian Industry Group & Environment Australia, 2000), http://natural-resources.org/minerals/generalforum/csr/docs/guidelines/Australian%20PER%20Framework.pdf, viewed 14 August, 2004.

Environment Australia, *Product Innovation The Green Advantage, An Introduction to Design for Environment for Australian Business*, (Canberra, Australia: Environment Australia, 2001), http://www.deh.gov.au/industry/finance/publications/pubs/producersguide.pdf, viewed 16 December 2004.

Environment Protection Authority, *Profits from Cleaner Production A Self Help Tool for Small and Medium Businesses*, (Sydney: NSW Environment Protection Authority and Department of State and Regional Development, 2000b), p. 68.

Environment Protection Authority (EPA), *Who Cares about the Environment in 2000, environmental knowledge, attitudes, behaviour in NSW*, (Sydney, NSW: Environment Protection Authority, 2000a), p. 73.

Environment Protection Authority (EPA), *Who Cares about the Environment in 2003, a survey of NSW people's knowledge, attitudes and behaviour*, (Sydney, NSW: Environment Protection Authority, 2003), p. 83.

Environmental Protection Authority, *Environmental Auditing*, (Barton, Australian Capital Territory: The Agency, 1996), p. 64.

Environmental Protection Authority NSW (EPA NSW), *Corporate Environmental Reporting Why and How*, (Sydney: NSW Environment Protection Authority, 1997), p. 28.

R. Estes, *Corporate social accounting*, (New York: John Wiley & Sons, Inc., 1976), p. 166.

D. Esty and K. Samuel-Johnson, *Environment by number*, (New Haven, Connecticut: Yale University, 2001), www.ciesin.org/indicators/esi, viewed 8 May 2004.

European Sustainable and Responsible Investment Forum (Eurosif), *Socially Responsible Investment among European Institutional Investors*, (Paris: Eurosif, 2003), p. 72.

European Sustainable and Responsible Investment Forum (Eurosif), *SRI Toolkit*, (Paris: Eurosif, 2004), p. 39.

European Union, *Communication From The Commission On The Precautionary Principle COM (2000) 1*, (Brussels: OECD, 2000), http://europa.eu.int/comm/environment/docum/20001 en.htm, viewed 12 December 2004.

European Union, *Guidance on EIA, EIS Review*, (Luxembourg: 2001), http://europa.eu.int/comm/environment/eia/eia-guidelines/g-review-full-text.pdf, viewed 18 December 2004.

European Union, *The European Union at a Glance*, (Brussels, Belgium: 2004), http://europa.eu.int/abc/index en.htm, viewed 27 June 2004.

C. Evans, 'News analysis: Corporate social responsibility – Sustainability: The bottom line', *Accountancy*, 131 (2003) 1313–1328.

D. Evans and S. Ross, 'The Role of Life Cycle Assessment in Australia', *Australian Journal of Environmental Management*, 5 (3) (1998) 137–145.

D. Farrier, 'In Search of Real Criminal Law', in T. Bonyhady, *Environmental Protection and Legal Change*, (Sydney: Federation Press, 1992), pp. 79–124.

D. Farrier, 'Fragmented Law in Fragmented Landscapes: The Slow Evolution of Integrated Natural Resource Management Legislation in NSW', *Environmental and Planning Law Journal*, 19 (2) (2002) 89–108.

R. H. Fazio and M. P. Zanna, 'Direct Experience and Attitude – Behaviour Consistency', in L. Berkowitz, *Advances in Experimental Social Psychology*, (New York: Academic Press, 1981), pp. 162–202.

J. Fien and D. Tilbury, 'The Global Challenge of Sustainability', in D. Tilbury, R. B. Stevenson, J. Fien and D. Schreuder, *Education and Sustainability: Responding to Global Change*, (IUCN: Gland, 2002), pp. 1–12.

F. Figge, T. Hahn, S. Schaltegger and M. Wagner, 'The Sustainability Balanced Scorecard – linking sustainability management to business strategy', *Business Strategy and the Environment*, 11 (5) (2002) 269–284.

B. Flyvbjerg, *Rationality and Power – Democracy in Practice*, (Chicago: The University of Chicago Press, 1998), p. 304.

X. Font and J. Tribe, 'Promoting green tourism: the future of environmental awards', *The International Journal of Tourism Research*, 3 (1) (2001) 9–21.

E. H. Foreman and M. A. Selly, *Decision by Objectives*, (Arlington, Va.: World Scientific, circa 2004), http://www.expertchoice.com/, viewed 14 March 2004.

M. Fowler, C. Hart and C. Phillips, 'Social and Environmental Reporting: A Snapshot of New Zealand', *Australian CPA*, 69 (11) (1999) 302.

R. J. Fowler, 'Environmental Law and its Administration in Australia', *Environmental and Planning Law Journal*, 1 (1) (1984) 10–49.

W. Fox, *Toward a Transpersonal Ecology: developing new foundations for environmentalism*, (New York: SUNY Press, 1995), p. 380.

C. Frankel, *In Earth's Company, Business, Environment and the Challenge of Sustainability*, (Gabriola Island, Canada: New Society Publishers, 1998), p. 223.

D. Freestone, 'New Directions in Environmental Law and Policy in the European Union', in B. Boer, R. Fowler and N. Gunningham, *Environmental Outlook No 2: Law and Policy*, (Sydney: Federation Press, 1996), pp. 29–52.

D. Frey, 'How green is BP', *The New York Times*, Dec 8 (2002) 1–7.

A. L. Friedman and S. Miles, 'Socially responsible investment and corporate social and environmental reporting in the UK: An exploratory study', *British Accounting Review*, 33 (4) (2001) 523–548.

M. Friedman, 'The Social Responsibility of Business is to Increase its Profits', *The New York Times Magazine*, September 13 (1970), pp. 32–33, pp. 122–126.

D. Frost, 'Don't just say you're green – prove it', *American Demographics*, 13 (6) (1991) 20–21.

D. A. Fuller, *Sustainable Marketing: Managerial-Ecological Issues*, (Thousand Oaks, CA: Sage Publications, 1999), p. 400.

L. Fulop and S. Linstead, *Management a Critical Text*, (Melbourne: Macmillan Education Australia, 1999), p. 472.

C. Fussler, *Sustainable Development*, World Business Council for Sustainable Development, 2004), http://www.wbcsd.org/plugins/DocSearch/details.asp?type=DocDet&DocId=MjU1Mg#, viewed 16 May 2004.

A. Gale, 'Coroner finds ESSO to blame for Longford gas disaster', *Financial Times*, Nov 15 (2002) 10.

S. M. Gardiner, 'The global warming tragedy and the dangerous illusion of the Kyoto Protocol', *Ethics & International Affairs*, 18 (1) (2004) 23–41.

L. Gettler, 'Green group seeks bank's help in Gunns war', *The Age*, July 10 (2002) 3.

D. Gibbs, 'Ecological modernisation, regional economic development and regional development agencies', *Geoforum*, 31 (2000) 9–19.

K. Gibson and R. Kendall, 'Australian Securities & Investment Commission and socially responsible investing', *Journal of the Asia Pacific Centre for Environmental Accountability*, 9 (1) (2003) 10–11.

A. Giddens, *The Consequences of Modernity*, (Stanford: Stanford University Press, 1990), p. 186.

A. Gilmour, *Personnel communication concerning lectures in Environmental Management 1 at the Graduate School of the Environment*, (Sydney: Macquarie University, 2004).

R. Giuntini and T. Andel, 'Master the six R's of reverse logistics (Part 2)', *Transportation & Distribution*, 36 (3) (1995) 93–98.

J. Glasson, R. Therivel and A. Chadwick, *Introduction to environmental impact assessment : principles and procedures, process, practice and prospects*, (London: UCL Press, 1999), p. 496.

J. Glenn, 'Keeping the Defense Department out of the landfill', *BioCycle*, 37 (4) (1996) 46–47.

Global Conscience Conference, *Global Conscience Conference*, (Copenhagen: Danish Sustainability Conference, 2004), http://www.globalconscience.dk/indeng.htm, viewed 20 August 2004.

Global Reporting Initiative, *Sustainability reporting guidelines*, (Amsterdam: UNEP/GRI, 2002), http: www.globalreporting.org/guidelines/2002.asp, viewed 14 August 2004.

Global Reporting Initiative, *Global Reporting Initiative Web Site*, (Amsterdam: GRI, 2004), www.globalreporting.org, viewed 1 June 2004.

T. F. Golob and D. A. Hensher, 'Greenhouse Gas Emissions and Australian Commuters' Attitudes and Behavior concerning Abatement Policies and Personal Involvement', *Transportation Research: Part D: Transport and Environment*, 3 (1) (1998) 1–18.

T. B. Gooley, 'Is there hidden treasure in your packaging?', *Logistics Management*, 35 (12) (1996) 19–23.

J. Gordon, 'Conflicting Time-Scales: Politics, the Media, and the Environment', in T. S. Driver and G. P. Chapman, *Time-Scales and Environmental Change*, (London: Routledge, 1996), p. 275.

A. Gouldson and J. Murphy, *Regulatory Realities: The Implementation and Impact of Industrial Environmental Regulation*, (London: Earthscan, 1998), p. 192.

T. E. Graedel, 'The evolution of industrial ecology', *Environmental science and technology A Pages*, 34 (1) (2000) 28A–31A.

R. Grant and E. Papadakis, 'Transforming environmental governance in a 'laggard' state', *Environmental and Planning Law Journal*, 21 (2) (2004) 144–160.

R. Gray, J. Bebbington and D. Walters, *Accounting for the Environment*, (London: Paul Chapman, 1993), p. 348.

R. Gray, D. L. Owen and C. Adams, *Accounting and Accountability: Changes and Challenges in Corporate Social and Environmental Accounting*, (Europe: Prentice-Hall, 1996), p. 331.

D. Grayson and A. Hodges, *Corporate Social Opportunity*, (Sheffield: Greenleaf Publications, 2004), p. 390.

A. Griffiths and J. Petrick, 'Corporate Architectures for Sustainability', *International Journal of Operations and Production Management*, 21 (12) (2001) 1573–1585.

D. Grinlinton, 'The 'Environmental Era' and the emergence of 'Environmental Law' in Australia – A survey of Environmental Legislation and Litigation 1967–1987', *Environmental and Planning Law Journal*, 7 (2) (1990) 74–105.

T. Grundy, *Gurus on Business Strategy*, (London: Thorogood, 2003), p. 216.

R. Guerrette, 'Environmental Integrity and Corporate Responsibility', *Journal of Business Ethics*, 5 (5) (1986) 409–515.

N. Gunningham and D. Sinclair, 'Non-point pollution, voluntarism and policy failure: lessons for the Swan-Canning', *Environmental and Planning Law Journal*, 21 (2) (2004a) 93–104.

N. Gunningham and D. Sinclair, 'Curbing non-point pollution: lessons for the Swan-Canning', *Environmental and Planning Law Journal*, 21 (3) (2004b) 181–99.

J. Guthrie and T. Carlin, 'Corporate Social Responsibility: Corporations, Fund Managers and the Financial Services Reform Act in practice', *MGSM unpublished working paper*, (2004).

J. Guthrie and M. R. Mathews, 'Corporate social accounting in Australasia', *Research in Corporate Social Performance and Policy*, (1985) 251–277.

J. Guthrie and L. Parker, 'Corporate social disclosure practice: A comparative international analysis', *Advances in Public Interest Accounting*, 3 (1990) 159–175.

T. Gutowski, C. Murphy, D. Allen, D. Bauer, B. Bras, T. Piwonka, P. Sheng, J. Sutherland, D. Thurston and E. Wolff, 'Environmentally benign manufacturing: Observations from Japan, Europe and the United States', *Journal of Cleaner Production Article in Press* (on-line http://www.sciencedirect.com), 13 (1) (2005) 1–17.

M. Haigh, *Managed Investments, Managed Disclosures, Griffith Business School Working Paper*, (Brisbane: Griffith University, 2004).

M. Hain and C. Cocklin, 'The effectiveness of the courts in achieving the goals of environment protection legislation', *Environmental and Planning Law Journal*, 18 (3) (2001) 319–338.

M. Hajer, *Politics of Environmental Discourse: Ecological Modernization and the Policy Process*, (Oxford: Oxford University Press, 1995), p. 332.

C. M. Hall, 'Integrated Heritage Management: dealing with principles, conflict, trust and reconciling stakeholder differences', *Heritage Economics Conference*, (Canberra: Australian Heritage Commission, 2000) 173–182.

C. Hamilton, *Growth Fetish*, (Crows Nest Australia: Allen & Unwin, 2003), p. 262.

C. Hamilton and R. Denniss, *Tracking Wellbeing in Australia: The Genuine Progress Indicator 2000, Discussion Paper No. 35*, (Canberra: The Australia Institute, 2000), p. 9.

R. Harding, *Environmental Decision-Making The Role of Scientists, engineers and the public*, (Leichhardt, Sydney: The Federation Press, 1998), p. 366.

L. C. Harland, 'Environmental Management Systems and how they can work for you', *Environmental Engineering Society, National Environmental Engineering Conference*, Institution of Engineers Australia, 2003), pp. 354–359.

K. Harrison and W. Antweiler, 'Incentives for Pollution Abatement: Regulation, Regulatory Threats and Non-Governmental Pressures', *Journal of Policy Analysis and Management*, 22 (3) (2003) 361–382.

S. R. Harrop, 'From cartel to conservation and on to compassion: animal welfare and the International Whaling Commission', *Journal of International Wildlife Law & Policy*, 6 (1–2) (2003) 79–105.

S. Hart, 'A natural resource based view of the firm', *Academy of Management Review*, 20 (1995) 986–1014.

S. Hart, 'Beyond Greening', *Harvard Business Review*, Jan–Feb (1997) 66–76.

C. L. Hartman, E. R. Stafford and M. J. Polonsky, 'Green Alliances: Environmental Groups as Strategic Bridges to Other Stakeholders', in M. Charter and M. J. Polonsky, *Greener Marketing: A Global Perspective to Greening Marketing Practice, Second Edition*, (Sheffield, UK: Greenleaf Publishing, 1999), pp. 201–217.

P. Hass, 'Epistemic communities and international policy coordination', *International Organization*, 46 (1) (1992) 1–35.

P. Hawken, B. L. Lovins and L. H. Lovins, *Natural Capitalism, The Next Industrial Revolution*, (London: Earthscan, 2000), p. 396.

B. Hedberg (ed.), *Virtual Organisations and Beyond: Discover Imaginary Systems*, (New York: Wiley, 1997), p. 250.

Hewlett Packard, *Global Citizenship Report*, (Palo Alto, California: Hewlett Packard, 2003), www.hp.com, viewed 15 May 2004.

Hill & Knowlton, *CEOs on Corporate Reputation Influencers, Corporate Social Responsibility, and Board-Level Involvement*, (New York: Hill and Knowlton, 2002), http://www.hillandknowlton.com/common/file.php/pg/duck/crw/binaries/72/Survey Results 2002.pdf, viewed 19 November 2004.

T. Hill and R. Westbrook, 'SWOT analysis: It's time for a product recall', *Long Range Planning*, 30 (1) (1997) 46–52.

R. Hillary, 'Environmental auditing: concepts, methods and developments', *International Journal of Auditing*, 12 (1998) 71–78.

D. Hini and P. Gendall, 'The link between environmental attitudes and behaviour', *Marketing Bulletin*, 6 (1995) 22–32.

M. A. Hitt, B. W. Keats and S. M. DeMarie, 'Navigating in the new competitive landscape: Building strategic flexibility and competitive advantage in the 21st century', *Academy of Management*, 12 (4) (1998) 22–42.

C. L. Ho, ' Pooled Cars', *Architecture*, 89 (10) (2000) 76.

A. J. Hoffman, *From Heresy to Dogma: An Institutional History of Corporate Environmentalism*, (San Francisco: New Lexington Books, 1997), p. 287.

A. J. Hoffman, 'Institutional Evolution and Change: Environmentalism and the U.S. Chemical Industry', *Academy of Management Journal*, 42 (4) (1999) 351–372.

A. J. Hoffman, *Competitive Environmental Strategy*, (Washington D.C.: Island Press, 2000), p. 256.

A. J. Hoffman, *From Heresy to Dogma, An Institutional History of Corporate Environmentalism*, (Stanford, California: Stanford University Press, 2001), p. 287.

M. W. Holdgate, M. Kassas and G. F. White, 'World Environmental Trends Between 1972 and 1982', *Environmental Conservation*, 9 (1) (1982) 11–29.

L. Holland, 'Can the principle of the ecological footprint be applied to measure the environmental sustainability of business?', *Corporate Social Responsibility and Environmental Management*, 10 (4) (2003) 224–232.

R. Hooghiemstra, 'Corporate communication and impression management – New perspectives why companies engage in corporate social reporting', *Journal of Business Ethics*, 27 (1/2) (2000) 55–68.

D. Hortensius, L. Bergenhenegouwen, R. Gouwens and A. De Jong, 'Towards a Generic Model for Integrating Management Systems', *ISO Management Systems*, January–February (2004) 21–28.

R. Horwood, 'Data Management is the Key to Effective Sustainability Reporting', *National Environmental Conference*, (Brisbane, Australia: Environmental Engineering Society of the Institution of Engineers Australia, 2003), 246–252.

T. Howard, 'Liability of Directors for Environmental Crime: The anything-but-level playing field in Australia', *Environmental and Planning Law Journal*, 17 (4) (2000) 250–271.

F. Hubbard, *Strategic Management: Thinking, Analysis and Action*, (Frenches Forest, Sydney, Australia: Pearson Education Pty Limited, 2000), p. 355.

C. Hunt and E. Auster, 'Proactive environmental management: Avoiding the toxic trap', *Sloan Management Review*, 31 (2) (1990) 7–18.

A. Hutchinson and F. Hutchinson, *Environmental business management: sustainable development in the new millenium*, (New York: McGraw-Hill, 1997), p. 390.

C. Hutchinson, 'Integrating environmental policy with business strategy', *Long Range Planning*, 29 (1) (1996) 11–23.

IASB, *International Accounting Standards Board web site*, (London: IASB, 2004), www.iasb.org/, viewed 19 June 2004.

ILO, *Tripartite Declaration of Principles Concerning Multinational Enterprises and Social Policy*, (Geneva: International Labour Organisation', 1977), http://www.ilo.org/public/english/standards/norm/sources/mne.htm, viewed 15 November 2004.

INECE, *International Network for Environmental Compliance and Enforcement*, (Washington, DC.: INECE, 2004), http://www.inece.org/), viewed 3 December 2004.

R. Inglehart, *Culture Shift in Advanced Industrial Society*, (Princeton: Princeton University Press, 1990), p. 504.

Institute of Social and Ethical AccountAbility (ISEA), *AccountAbility 1000 (AA1000) framework: Standards, guidelines and professional qualification*, (London: ISEA, 1999), viewed 14 August 2004.

International Association of Public Participation (IAP2), *Certificate Course in Public Participation Module 1: The IAP2 Foundations of Public Participation*, (Denver, Colorado: IAP2, 2000a).

International Association of Public Participation (IAP2), *The IAP2 Public Participation Spectrum*, (Denver, Colorado: IAP2, 2000b), www.iap2.org, viewed 25 November 2004.

International Energy Agency, *World Energy Outlook 1998 Edition*, (Paris: International Energy Agency, 1998), http://library.iea.org/dbtw-wpd/textbase/nppdf/free/1998/weo98.pdf, viewed 13 July 2004.

International Energy Agency, *World Energy Outlook 2001 Insights*, (Paris: International Energy Agency, 2001), http://library.iea.org/dbtw-wpd/textbase/nppdf/free/1999/weo1999.pdf, viewed 16 July 2004.

International Energy Agency, *Key World Energy Statistics 2003*, (Paris: International Energy Agency, 2003), http://library.iea.org/dbtw-wpd/Textbase/nppdf/free/2003/key2003.pdf, viewed 16 July 2004.

International Organization for Standardization (ISO), *ISO 14010: 1996 (superseded) Guidelines for environmental auditing – General principles*, (Geneva: International Organization for Standardization, 1996), p. 6.

International Organization for Standardization (ISO), *ISO 14001: 1996 (superseded) Environmental management systems – Specification with guidance for use*, (Geneva: International Organization for Standardization, 1996a), p. 16.

International Organization for Standardization (ISO), *ISO 14004: 2004 (superseded) Environmental management systems – General guidelines on principles, systems and supporting techniques*, (Geneva: International Organization for Standardization, 1996b), p. 32.

International Organization for Standardization (ISO), *ISO 14040: 1997 Environmental management – Life cycle assessment – Principles and framework*, (Geneva: International Organization for Standardization, 1997), p. 12.

International Organization for Standardization (ISO), *ISO 14041: 1998 Environmental management – Life cycle assessment – Goal and scope definition and inventory analysis*, (Geneva: International Organization for Standardization, 1998), p. 22.

International Organization for Standardization (ISO), *ISO 14031: 1999 Environmental management – Environmental performance evaluation – Guidelines*, (Geneva: International Organization for Standardization, 1999a), p. 32.

International Organization for Standardization (ISO), *ISO 14032: 1999 Environmental management – Examples of environmental performance evaluation (EPE)*, (Geneva: International Organization for Standardization, 1999b), p. 93.

International Organization for Standardization (ISO), *ISO 14042: 2000 Environmental management – Life cycle assessment – Life cycle impact assessment*, (Geneva: International Organization for Standardization, 2000), p. 17.

International Organization for Standardization (ISO), *ISO 19011: 2002 Guidelines for quality and/or environmental management systems auditing*, (Geneva: International Organization for Standardization, 2002), p. 31.

International Organization for Standardization (ISO), *ISO 14001: 2004 Environmental management systems – Specification with guidance for use*, (Geneva: International Organization for Standardization, 2004), p. 14.

International Organization for Standardization (ISO), *International Organization for Standardization's web site*, (Geneva: International Organization for Standardization, 2004a), viewed 18 December 2004.

International Organization for Standardization (ISO), *ISO 14004: 2004 Environmental management systems – General guidelines on principles, systems and supporting techniques*, (Geneva: International Organization for Standardization, 2004b), p. 39.

IUCN, *ECOLEX, A Gateway to Environmental Law*, (2004), http://www.ecolex.org/, viewed 3 December 2004.

L. Jacobson, 'Green giants', *National Journal*, 25 (19) (1993) 1113–1116.

C. Jasch, 'Environmental performance evaluation and indicators', *Journal of Cleaner Production*, 8 (2000) 179–88.

L. Johannson, 'Caught in a crossfire? Balancing economic, environmental and social considerations while "in compliance"', *Total Quality Environmental Management*, 3 (3) (1994) 363–372.

G. Johnson and K. Scholes, *Exploring Corporate Strategy 6th Edition*, (Harlow, England: Pearson Education Limited, 2002), p. 1082.

M. Johnson, 'Regulating the Environment', in B. Head and E. McCoy, *Deregulation or Better Regulation*, (Brisbane: Macmillan, 1991), pp. 131–140.

A. Jordan, Wurzel, A. Zito and L. Brückner, 'Policy Innovation or "Muddling Through"? "New" Environmental Policy Instruments in the United Kingdom', *Environmental Politics*, 12 (2003b) 179–197.

A. Jordan, R. Wurzel, A. Zito and L. Brückner, 'The Innovation and Diffusion of "New" Environmental Policy Instruments (NEPIs) in the European Union and its Member States', *Proceedings of the 2001 Berlin Conference on the Human Dimensions of Global Environmental Change, Global Environmental Change and the Nation State*, (Potsdam: Potsdam Institute for Climate Impact Research, 2002).

A. Jordan, R. Wurzel and A. R. Zito, '"New" Environmental Policy Instruments: An Evolution or a Revolution in Environmental Policy?' *Environmental Politics*, 12 (2003a) 201–224.

B. E. Joyner and D. Payne, 'Evolution and implementation: A study of values, business ethics and corporate social responsibility', *Journal of Business Ethics*, 41 (4) (2002) 297–311.

L. S. Kane, 'How can we stop corporate environmental pollution: corporate officer liabilities', *New England Law Review*, 26 (1991) 293–317.

N. Kangun and M. J. Polonsky, 'Regulation of Environmental Marketing Claims: A Comparative Perspective', *International Journal of Advertising*, 11 (1) (1995) 1–24.

N. Kanie and P. Hass, *Emerging Forces in Environmental Governance*, (Tokyo: United Nations University Press, 2004), p. 380.

I. Kant, *Lectures on ethics (translated by Louis Infield)*, (New York: Harper & Row, 1963), p. 253.

D. Kapelianis and S. Strachan, 'The Price Premium of an Environmentally Friendly Product', *South African Journal of Business Management*, 27 (4) (1996) 89–95.

R. S. Kaplan and D. P. Norton, 'The Balanced Scorecard – Measures That Drive Performance', *Harvard Business Review*, 70 (1) (1992) 71–80.

R. Kemp, 'Communicating the Results of a Risk Assessment: Lessons from Radioactive Waste Disposal', in M. Quint, D. Taylor and R. Purchase, *Environmental Impact of Chemicals: Assessment and Control*, (London: Royal Society of Chemistry, 1996), pp. 209–222.

T. Kemp, *Personal communication on Balanced Scorecard implementation*, (Sydney: Thompson Legal and Regulatory Asia Pacific Corporation, 2004).

G. A. Keoleian, D. Menerey, B. W. Vigon, D. A. Tolle, B. W. Cornaby, H. C. Latham, C. L. Harrison, T. L. Boguski, R. G. Hunt and J. D. Sellers, *Product Life Cycle Assessment to Reduce Health Risks and Environmental Impacts*, (New Jersey, USA: Noyes Publications, 1994), p. 288.

W. E. Kilbourne, 'Green Advertising: Salvation or Oxymoron?', *Journal of Advertising*, 24 (2) (1995) 7–20.

W. E. Kilbourne and S. Beckman, 'Review and Critical Assessment of Research on Marketing and the Environment', *Journal of Marketing Management*, 14 (6) (1998) 513–532.

W. E. Kilbourne, S. Beckman and E. Thelen, 'The role of the Dominate Social Paradigm in Environmental Attitudes: A Multi-National Examination', *Journal of Business Research*, 55 (3) (2002) 193–204.

A. King, 'Avoiding ecological surprise: Lessons from long standing communities', *Academy of Management Review*, 20 (4) (1995) 961–985.

A. King, 'How to get started in corporate social responsibility', *Financial Management*, October (2002) 5.

F. Kingham, 'The Impact on Environmental Practice of the Environmental Protection Legislation (QLD)', *Australian Environmental Law News*, 3 (1995) 62–72.

T. Kletz, 'Lessons from Longford – The Esso gas plant explosion', *Chemical Engineering Progress*, 97 (9) (2001) 78–80.

C. Knill and A. Lenschow, '"New" Environmental Policy Instruments as a Panacea? Their Limitations in Theory and Practice', in K. Holzinger and P. Knoepfel, *Environmental Policy in a European Community of Variable Geometry: The Challenge of the Next Enlargement*, (Base: Helbing and Lichtenhahn, 2000), pp. 317–348.

P. Knudtson and D. Suzuki, *Wisdom of the Elders*, (North Sydney: Allen & Unwin, 1992), p. 232.

P. Kok, T. Wiele, R. McKenna and A. Brown, 'A corporate social responsibility audit within a quality management framework', *Journal of Business Ethics*, 31 (4) (2001) 285–297.

A. Kolk and A. Mauser, 'The evolution of environmental management: from stage models to performance evaluation', *Business Strategy and the Environment*, 11 (1) (2001) 14–31.

D. Korten, *When Corporations Rule the World*, (London: Earthscan, 1995), p. 374.

P. Kotler, *Marketing Management 11th ed.*, (USA: Prentice Hall, 2003), p. 706.

KPMG, *UK Survey of Environmental Reporting*, (London, UK: KPMG, 1997).

KPMG International Environment Network and Institute for Environment Management (University of Amsterdam), *KPMG International Survey of Environmental Reporting 1999*, (The Netherlands: KPMG Environmental Consulting, 1999), viewed 22 December 2004.

J. E. Krier, 'The Tragedy of the Commons, Part Two', *Harvard Journal of Law and Public Policy*, 15 (1992) 325–347.

P. Lal and E. Young, 'The role and relevance of Indigenous cultural capital in environment management in Australia and the Pacific', *Heritage Economics Conference, Challenges for heritage conservation and sustainable development in the 21st century*, (Canberra: Australian Heritage Commission, 2000) 195–217.

R. Lamming, A. Faruk and P. Cousins, 'Environmental Soundness: A Pragmatic Alternative to Expectations of Sustainable Development in Business Strategy', *Business Strategy and the Environment*, 8 (1999) 177–188.

K. Lapthorne, 'Esso at fault over fatal blast', *Herald Sun*, Nov. 16 (2002) 6.

T. Latvala and J. Kola, 'Consumers Willingness to Pay for Information about Food Safety and Quality: Case Beef', *IAMA World Food and Agribusiness Conference June 24–28*, (Chicago IL: IAMA, 2000) 7.

G. Leane, 'Environmental Law's Liberal Roots: (Not) a Green Paradigm', in N. Rogers, *Green Paradigms and the Law*, (Lismore: Southern Cross University Press, 1998), pp. 1–31.

D. Lee, P. Newman and R. Price, *Decision Making in Organisations*, (London: Financial Times Pitman Publishing, 1999), p. 262.

A. Leopold, *A Sand County Almanac and sketches here and there*, (New York: Oxford University Press, 1987, c1949), p. 228.

G. Levine, 'Green marketing gets cautious', *Advertising Age*, 64 (28) (1993) 4.

M. Levine, *Threats to the Environment*, (Melbourne: Roy Morgan Research, 2002), www.roymorgan.com/news/polls/2002/, viewed 1 July 2004.

A. Lewis, 'Disposal disorder: automakers lead in recycling while the European Commission lags behind in writing end-of-life rules. But the whole industry wants to know who will pay for it,' *Automotive Industries*, 182 (8) (2002) 28–31.

D. F. Linowes, 'An Approach to Socio-Economic Accounting', *The Conference Board RECORD*, Nov. (1972) 58–61.

S. Lloyd, 'Greedy, Dishonest, Boring, Faceless', *Business Review Weekly*, April 22–28 (26) (2004) 32–39.

D. J. Lober, 'Explaining the Formation of Business-Environmentalist Collaborations: Collaborative Windows and the Paper Task Force', *Policy Sciences*, 30 (1) (1997) 1–24.

J. Loh, *WWF Living Planet Report 2002*, (Cambridge, UK: WWF-World Wide Fund For Nature, 2002), p. 37.

J. R. Lothian, 'Attitudes of Australians Towards the Environment: 1975 to 1994', *Australian Journal of Environmental Management*, 1 (2) (1994) 78–99.

J. R. Lothian, 'Australians Attitudes Towards the Environment: 1991 to 2001', *Australian Journal of Environmental Management*, 9 (1) (2002) 45–62.

A. Lovins, L. Lovins and P. Hawken, 'A roadmap for natural capitalism', *Harvard Business Review*, 77 (3) (1999) 145–158.

L. Lovins and A. Lovins, 'Pathway to sustainability', *Forum for Applied Research and Public Policy*, 15 (4) (2000) 13–22.

I. Lowe, 'The Greenhouse Effect and the Politics of Long-Term Issues', in S. Bell and B. Head, *State, Economy and Public Policy in Australia*, (Oxford: Oxford University Press, 1994), p. 400.

N. Luttbeg, *Public Opinion and Public Policy: Models of Political Linkage*, (Homewood, Illinois: The Dorsey Press, 1968), p. 469.

H. Mackay, *Why don't people listen: Solving the communication problem*, (Sydney: Pan Macmillan Publishers, 1994), p. 352.

J. Macken, 'Trick or treat', *Australian Financial Review*, Oct. 11 (2002) 36.

J. Madden, 'Oil giant caused fatal gas disaster', *The Weekend Australian*, Nov 16 (2002) 4.

T. G. Marx, 'Corporate social performance reporting', *Public Relations Quarterly*, 37 (4) (1992/93) 38–44.

J. Mathews, *TCG: Sustainable Economic Organisation Through Networking UNSW Studies in Organisational Analysis and Innovation*, (Sydney: Industrial Relations Research Centre, University of New South Wales, 1992).

P. Matson, 'Environmental Challenges for the Twenty-First Century: Interacting Challenges and Integrative Solutions', *Ecology Law Quarterly*, 27 (4) (2001) 1179–1190.

S. Mattieu, 'Integrated Management Systems', *Management Systems The International Review of ISO 9000 and ISO 14000*, 3 (4) (2003) 39–44.

L. McGeachy, 'Trends in Magazine Coverage of Environmental Issues', *Journal of Environmental Education*, 20 (2) (1989) 6–13.

D. H. Meadows, D. L. Meadows and J. Randers, *Beyond the Limits: Confronting Global Collapse Envisioning a Sustainable Future*, (Post Mills, Vermont: Chelsea Green Publishing, 1992), p. 300.

D. H. Meadows, D. L. Meadows, J. Randers and W. W. Behrens, *The Limits to Growth*, (Washington, D.C.: Universe Books, Potomac Associates Chelsea Green Publishing, 1972), p. 205.

A. L. Meijnders, C. J. H. Midden and H. A. M. Wilke, 'Communications About Environmental Risks and Risk-Reducing Behavior: The Impact of Fear on Information Processing', *Journal of Applied Social Psychology*, 31 (4) (2001) 754–777.

N. Mendleson and M. J. Polonsky, 'Using Strategic Alliances to Develop Credible Green Marketing', *Journal of Consumer Marketing*, 12 (2) (1995) 4–18.

A. Menon and A. Menon, 'Enviropreneurial marketing strategy: The emergence of corporate environmentalism as market strategy', *Journal of Marketing*, 61 (1) (1997) 51–67.

D. Mercer, *'A Question of Balance': Natural Resources Conflict Issues in Australia*, (Sydney: Federation Press, 1991), p. 346.

H. W. Micklitz, 'The German Packaging Order? A Model for State? Induced Waste Avoidance?' *The Columbia Journal of World Business*, September (1992) 120–127.

L. W. Milbraith, 'Culture and the Environment in the United States', *Environmental Management*, 9 (2) (1985) 161–172.

L. Milbrath, *Envisioning a Sustainable Society: Learning Our Way Out*, (New York: State University of New York Press, 1989), p. 403.

M. Miller, *The Third World in Global Environmental Politics*, (Boulder: Lynne Rienner, 1995), p. 181.

D. M. Minoli and J. N. B. Bell, 'Insurance as an alternative environmental regulator: findings from a retrospective pollution claims survey', *Business Strategy and the Environment*, 12 (2) (2003) 107–117.

R. C. Mitchell, A. G. Mertig and R. E. Dunlap, 'Twenty years of Environmental Mobilisation: Trends among National Environmental Organisations', *Society and Natural Resources*, 4 (1991) 219–234.

K. Miyamoto, 'Japanese Environmental Policies Since World War II', *Capitalism, Nature, Socialism*, 2 (2) (1991) 1–9.

L. A. Mohr, D. Eroglu and P. S. Ellen, 'The Development and Testing of a Measure of Scepticism towards Environmental Claims in Marketers' Communications', *The Journal of Consumer Affairs*, 32 (1) (1998) 30–56.

L. A. Morris, M. Hastak and M. B. Mazis, 'Consumer Comprehension of Environmental Advertising and Labeling Claims', *Journal of Consumer Affairs*, 22 (4) (1995) 439–460.

P. Morris and P. Therivel, *Methods of environmental impact assessment*, (London: Shon Press, 2001), p. 378.

T. Mottershead (ed.), *Environmental Law and Enforcement in the Asia-Pacific Rim*, (Hong Kong: Sweet and Maxwell Asia, 2002), p. 649.

A. E. Mudge, *Value Engineering A Systematic Approach*, (New York: McGraw-Hill Book Company, 1971), p. 286.

A. Naess, 'The shallow and the deep, long-range ecology movements: a summary', *Inquiry*, 16 (1973) 95–100.

B. Nattrass and M. Altomare, *The Natural Step for Business*, (Gabriola Island: New Society Publishers, 1999), p. 240.

Nature Conservation Council, *Green Games Watch 2000*, (Sydney: Nature Conservation Council of NSW, Australia, 2001), http://www.nccnsw.org.au/member/ggw/about/evo.html, viewed 22 September 2004.

S. J. Newell, R. E. Goldsmith and E. J. Banzhaf, 'The Effect of Misleading Environmental Claims on Consumer Perceptions of Advertisements', *Journal of Marketing Theory and Practice*, Spring (1998) 48–60.

W. Nicholson, *Microeconomic Theory, Basic Principles and Extensions (6 ed.)*, (Sydney, Australia.: Dryden Press, 1995), p. 901.

W. Noble and I. Davidson, *Human Evolution, Language and Mind*, (Cambridge: Cambridge University Press, 1997), p. 272.

N. Nohria and R. Eccles, *Beyond the Hype: Rediscovering the Essence of Management*, (Boston: Harvard Business School Press, 1992), p. 278.

B. Norton, 'Sustainability, human welfare and ecosystem health', *Environmental Values*, 1 (1992) 97–111.

B. Norton, 'Why I am not a non-anthropocentrist: Callicott and the failure of monistic inherentism', *Environmental Ethics*, 17 (1995) 341–458.

NSW Government, *Participation and the NSW Policy Process, A Discussion Paper for the Cabinet Office*, (Sydney: New South Wales State Government, 1998).

NSW Roads and Traffic Authority (RTA), *Community Involvement Practice Notes and Resource Manual*, (Sydney: NSW State Government, 1998), www.rta.nsw.au/ environment/downloads/cirmanual d/1.html, viewed 28 December 2004.

B. T. Oakley, 'Total Quality Product Design – How to Integrate Environmental Criteria into the Production Realization Process', *Total Quality Environmental Management*, 2 (3) (1993) 309–321.

OECD, *Guidelines for Multinational Enterprises*, (Paris: Organisation for Economic Co-operation and Development, 2000), http://www.oecd.org/department/ 0,2688,en 2649 34889 1 1 1 1 1,00.html, viewed 15 November 2004.

OECD, *Environmentally related taxes in OECD countries, issues and strategies*, (Paris: Organisation for Economic Co-operation and Development, 2001), p. 100.

OECD, *Discussion paper for Conference on Environmental Fiscal Reform, 27 June 2002*, (Berlin: Organisation for Economic Co-operation and Development, 2002), p. 17.

OECD, *Tradeable permits: Policy Evaluation Design and Reform*, (Paris: Organisation for Economic Co-operation and Development, 2004a), p. 192.

OECD, *Environmentally Related Taxes database*, (Paris: Organisation for Economic Co-operation and Development, 2004b), www.oecd.org/env/tax-database, viewed 24 October 2004.

C. Olivier, 'Tapping into the green dollar', *Corporate Finance*, 201 (2001) 39–40.

Olympic Co-ordination Authority (OCA), *State of the Environment*, (Sydney: OCA, 1996), p. 104.

Olympic Co-ordination Authority (OCA), *State of the Environment*, (Sydney: OCA, 1997), p. 119.

L. Onisto, 'The business of sustainability', *Ecological Economics*, 29 (1) (1999) 39–45.

F. Orecchini and D. Sabatini, 'Cars and the environment: a new approach to assessment through ISO 14001 certification of the car process', *Proceedings of the Institution of Mechanical Engineers – Part D – Journal of Automobile Engineering*, 217 (1) (2003) 31–41.

T. O'Riordan, 'Frameworks for Choice: Core Beliefs and the Environment', *Environmental and Planning Law Journal*, 37 (8) (1995) 4–9, 25–9.

T. O'Riordan and H. Voisey, *The Transition to Sustainability: The Politics of Agenda 21 in Europe*, (London: Earthscan, 1998), p. 320.

S. Oskamp, 'Resource conservation and recycling: behavior and policy', *Journal of Social Issues*, 51 (4) (1995) 157–178.

J. Ottman, 'Source reduction will be more valuable in years ahead', *Marketing News*, 27 (23) (1993) 10.

J. A. Ottman, *Green Marketing: Opportunity for Innovation 2nd ed.*, (Chicago: NTC Business Books, 1998), p. 270.

M. Overcash, 'The evolution of US pollution prevention, 1976–2001: a unique chemical engineering contribution to the environment – a review', *Journal of Chemical Technology & Biotechnology*, 77 (11) (2002) 1197–1205.

R. Paden, 'Moral metaphysics, moral revolutions and environmental ethics', *Agriculture and Human Values*, Summer-Fall Edition (1990) 70–78.

R. J. Pagan, P. Prasad, J. C. Diniz da Costa and R. Van Berkel, 'Cleaner Production Applications for Coal Utilization from National Environment Conference', (Brisbane, Australia: Institution of Engineers, 2003) 372–378.

T. Paine, *Rights of Man (1792, reprinted 2004)*, (Adelaide: University of Adelaide Library Electronic Texts Collection, 2004), http://etext.library.adelaide.edu.au/p/p147r/, viewed 25 November 2004.

M. Painter, 'Participation and Power', in M. Munro-Clark, *Citizen Participation in Government*, (Sydney: Hale & Ironmonger Publishers, 1992), pp. 32–35.

G. Palast, *Power outage traced to dim bulb in White House: The tale of the Brits who swiped 800 jobs from New York, carted off $90 million, then tonight turned off our lights*, (London: Greg Palast, 2003), www.GregPalast.com, viewed 19 August 2004.

I. Palmer and R. Dunford, 'Organizing for hyper-competition: new organizational forms for a new age?', *New Zealand Strategic Management*, Summer (1997) 38–45.

E. Papadakis, *Politics and the Environment*, (Sydney: Allen and Unwin, 1993), p. 232.

R. Parker, 'Fund move gives buyers chance to take Initiative', *Supply Management*, April 26 (2001) 9.

J. W. Parlour and S. Schatzow, 'The Mass Media and Public Concern for Environmental Problems in Canada', *International Journal of Environmental Studies*, 13 (1978) 9–17.

T. M. Parris, 'Exploring Attitudes toward the Environment', *Environment*, 45 (6) (2003) 3.

J. Passmore, *Man's Responsibility for Nature: ecological problems and Western traditions*, (London: Duckworth, 1974), p. 213.

K. Peattie, 'Trappings versus substance in the greening of marketing planning', *Journal of Strategic Marketing*, 7 (1999) 131–148.

K. Peattie, 'Golden goose or wild goose? The hunt for the green consumer', *Business Strategy and the Environment*, 10 (4) (2001) 187–199.

F. B. Pedersen, *Ramboll Holistic Accounting and Capitalisation*, (Copenhagen, Denmark: Ramboll, 1999), http://www.ramboll.dk/docs/dan/Pressecenter/Publikationer/generelle/capitalization.pdf, viewed 8 June 2004.

R. Peglau, *The Number of ISO 14000 Registrations Throughout the World*, (Japan: Federal Environment Agency and ISO World, 2004), http://www.ecology.or.jp/isoworld/english/analy14k.htm, viewed 1 June 2004.

J. Persico Jr, *The TQM Transformation, A Model for Organisational Change*, (New York: Quality Resources, 1992), p. 195.

J. Pfeffer and J. Veiga, 'Putting People First for Organisational Success', *Academy of Management Executive*, 13 (1999) 37–48.

B. Pheasant, 'Esso and BHP under scrutiny', *Australian Financial Review*, Oct 14 (1998) 4.

B. Piasecki, *Corporate Environmental Strategy: the avalanche of change since Bhopal*, (New York: Wiley, 1995), p. 180.

A. Pike, 'When life is more than just a lottery: Management social reporting: A consultation exercise will help to ensure Camelot maintains closer links with employees, shareholders and the public', *Financial Times*, April 12 (2000) 18.

M. J. Polonsky, 'Green Marketing', in M. Charter and U. Tischner, *Sustainable Solutions: Sustainable and Eco-Products and Services*, (Sheffield, UK: Greenleaf Publishing, 2001b), pp. 282–303.

M. J. Polonsky, L. Carlson, S. Grove and N. Kangun, 'International Environmental Marketing Claims: Real Changes or Simple Posturing?', *International Marketing Review*, 14 (4) (1997) 218–232.

M. J. Polonsky and J. Ottman, 'Stakeholders Contribution to the Green New Product Development Process', *Journal of Marketing Management*, 14 (6) (1998) 533–558.

M. J. Polonsky and P. J. Rosenberger III, 'Re-evaluating to Green Marketing – An Integrated Approach', *Business Horizons*, 44 (5) (2001a) 21–30.

M. E. Porter, 'How competitive forces shape strategy', *Harvard Business Review*, 57 (2) (1979) 137–145.

M. E. Porter and C. van der Linde, 'Green and Competitive: Ending the Stalemate', *Harvard Business Review*, 73 (5) (1995) 120–133.

D. Potts, 'Green groups put pressure on bank', *Sun Herald*, Oct 6 (2002) 3.

L. Preston, 'Sustainability at Hewlett-Packard: From Theory to Practice', *California Management Review*, 43 (3) (2001) 26–38.

Public Accounts and Estimates Committee, *Issues paper no.3: Environmental Accounting and Reporting*, (Melbourne, Australia: Parliament of Victoria, 1998).

J. Quinn, *Intelligent Enterprise*, (New York: Free Press, 1992), p. 473.

D. W. Rajecki, *Attitudes*, (Sunderland, Massachusetts: Sinauer Associates Inc. Publishers, 1990), p. 522.

T. Regan, *Earthbound: new introductory essays in environmental ethics*, (New York: Random House, 1984), p. 371.

T. Regan and P. Singer, *Animal Rights and Human Obligations*, (Englewood Cliffs, New Jersey: Prentice Hall, 1989), p. 280.

Regulatory Impact Unit, *Better Policy Making A Guide to Regulatory Impact Assessment*, (London: Cabinet Office, 2003), http://www.cabinet-office.gov.uk/regulation/docs/ria/pdf/ria-guidance.pdf, viewed 18 June 2004.

R. D. Reynnells, 'Turning animal by-products into resource', *Biocycle*, 40 (6) (1999) 48, 51.

B. J. Richardson, 'Environmental Regulation through Financial Institutions: New Pathways for Disseminating Environmental Policy', *Environmental and Planning Law Journal*, 19 (1) (2002) 58–77.

N. Robinson, 'The 104th United States Congress: Environmental Law at the Precipice?' in B. Boer, R. Fowler and N. Gunningham, *Environmental Outlook No 2: Law and Policy*, (Sydney: Federation Press, 1996), pp. 18–28.

B. Rollin, 'Intrinsic value for nature: an incoherent basis for environmental concern', *Free Inquiry*, 13 (2) (1993) 20–22.

N. Roome, 'Developing environmental management systems', *Business Strategy and the Environment*, 1 (1992) 11–24.

D. B. Rose, *Nourishing Terrains: Australian Aboriginal Views of Landscape and Wilderness*, (Canberra, Australia: Australian Heritage Commission, 1996), p. 95.

W. A. Rosenbaum, *Environmental Politics and Policy 2nd Ed.*, (Washington, D.C.: CQ Press, 1991), p. 336.

A. Ross-Smith, 'Actioning Corporate Sustainability', in S. Benn, D. Dunphy, A. Griffiths and A. Ross-Smith, *All-Academy Symposium at Academy of Management Conference, August 2004*, (New Orleans: 2004).

A. Roy, *The algebra of infinite justice*, (London: Flamingo, 2002), p. 324.

R. Roy, 'Design and Marketing Greener Products: the Hoover Case', in M. Charter and M. J. Polonsky, *Greener Marketing (2nd Edition)*, (Sheffield: Greenleaf Publishing, 1999), pp. 126–144.

M. V. Russo, *Environmental Management Readings and Cases*, (Boston: Houghton Mifflin Company, 1999), p. 431.

P. Ryan, 'Metropolitan Planning in Australia: The Instruments of Planning – Regulation', *Environmental and Planning Law Journal*, 5 (2) (1988) 147–158.

P. Ryan, 'Did we? Should we? Revisiting the 70s' Environmental Law Challenge in NSW', *Environmental and Planning Law Journal*, 18 (6) (2001) 561–578.

P. Ryan, 'Sustainability partnerships: eco-strategy theory in practice?' *Management of Environmental Quality: An International Journal*, 14 (2) (2003) 256–278.

P. Ryan, 'Internet marketing standards: Institutional coherence issues', *International Journal of Internet Marketing and Advertising*, 1 (1) (2004a) 85–103.

P. Ryan and N. Wayuparb, 'Greenspace Sustainability in Thailand', *Sustainable Development*, 12 (2004b) 223–237.

S. C. Johnson, *U.S. EPA Honors SC Johnson with First-Ever Environmental Award*, (Racine, Wisconsin: SC Johnson, 2004), http://www.scjohnson.com/family/fam_pre_pre_news.asp?art_id=95, viewed 1 July 2004.

S. Sagawa and E. Segal, 'Common Interest, Common Good: Creating Value Through Business and Social Sector Partnerships', *California Management Review*, 42 (2) (2000) 105–123.

R. Schaffer, 'Establishing the elements of an environmentally-aware corporate culture', *Environmental Planning and Law Journal*, 9 (1) (1992) 44–50.

S. Schaltegger and R. Burritt, *Contemporary Environmental Accounting: Issues, Concepts and Practice*, (United Kingdom: Greenleaf Publishing, 2000), p. 462.

S. Schaltegger, R. Burritt and H. Petersen, *An Introduction to Corporate Environmental Management*, (United Kingdom: Greenleaf Publishing, 2003), p. 384.

B. D. Schegelmilch, A. Diamantopoulos and G. M. Bohlen, 'Environmental Issues in the Freight Transport industry: A Qualitative Analysis of Key Stakeholders' Perceptions', in M. J. Polonsky and A. T. Mintu-Wimsatt, *Environmental Marketing: Strategies, Practice, theory and Research*, (New York: The Haworth Press, 1995), pp. 363–388.

S. Schmidheiny, *Changing Course: A Global Perspective on Development and the Environment*, (Cambridge, MA: The MIT Press, 1992), p. 374.

D. Schon, *Beyond the Stable State*, (New York: Random House, 1971), p. 254.

R. Schwarze, 'Environmental Liability and Accident Prevention: Preliminary Experiences in Germany', *European Environment*, 11 (2001) 314–323.

P. Senge and S. Carstedt, 'Innovating our way to the next industrial revolution', *Sloan Management Review*, Winter (2001) 24–38.

M. Sharfman, R. T. Ellington and M. Meo, 'The Next Step in Becoming "Green": Life-Cycle Oriented Environmental Management', *Business Horizons*, 40 (3) (1997) 13–22.

S. Sharma and H. Vredenburg, 'Proactive corporate environmental strategy and the development of competitively valuable organisational capabilities', *Strategic Management Journal*, 19 (1998) 729–753.

P. Shrivastava, 'The role of corporations in achieving ecological sustainability', *Academy of Management Review*, 20 (4) (1995) 936–960.

L. J. Shrum, J. A. McCarty and T. M. Lowrey, 'Buyer Characteristics of the Green Consumer and Their Implications for Advertising Strategy', *Journal of Advertising*, 24 (1995) 21–31.

J. Simms, 'Business: Corporate social responibility – You know it makes sense', *Accountancy*, 130 (1311) (2002) 48–50.

S. Simon and J. Proops, *Greening the Accounts*, (Cheltenham, United Kingdom: Edward Elgar Publishing Inc, 2000), p. 261.

J. A. Sinden and D. J. Thampapillai, *Introduction to Benefit-Cost Analysis*, (Melbourne: Longman Australia, 1995), p. 262.

S. F. Singer, 'Sustainable development vs. global environment: resolving the conflict', *Columbia Journal of World Business*, 27 (3–4) (1992) 154–163.

R. Singleton, P. Castle and D. Short, *Environmental Assessment*, (London: Thomas Telford Publishing, 1999), p. 288.

A. Skorecki, 'A Footsie Index for corporate ethics: The launch this summer of FTSE4 Good will put companies' standards in the spotlight', *Financial Times*, April 27 (2001a) 18.

A. Skorecki, 'Top companies under pressure on environment', *Financial Times*, April 10 (2001b) 23.

J. Slavich, 'Beyond permitting: Environmental compliance management challenges for the 1990s', *Journal of Environmental Permitting*, 2 (4) (1993) 525–541.

P. Slovic, 'Beyond Numbers: A Broader Perspective on Risk Perception and Risk Communication', in D. Mayo and R. Hollander, *Acceptable Evidence: Science and Values in Risk Management*, (New York: Oxford University Press, 1991), pp. 48–65.

M. S. Soroos, 'Trends in the Perception of Ecological Problems in the United Nations General Debates', *Human Ecology*, 9 (1) (1981) 23–45.

R. Sparkes, *Socially Responsible Investment – A Practical Guide for Professional Investors*, (Canada: John Wiley and Sons, 2002), p. 350.

D. Stace and D. Dunphy, *Beyond the Boundaries*, (Roseville: McGraw Hill, 2001), p. 288.

R. Staib, *Audit Reports of Design Consultants and Construction Contractors for the Rouse Hill trunk water infrastructure project – unpublished reports*, (Sydney: Rouse Hill Infrastructure Consortium Pty Ltd, 1993–2004), p. various.

R. Staib, *Solving Major Pollution Problems: A New Process Model*, unpublished PhD thesis, (Sydney: Macquarie University, 1997), p. 200.

R. Staib, *Audit Reports of Design Consultants and Construction Contractors for the Homebush Bay Infrastructure Works – Sydney 2000 Olympics – unpublished reports*, (Sydney: Consultant GHD Pty Ltd, 1997–2000), p. various.

R. Staib, 'Processes in Pollution Management: An Australian Model', *Environmental Management*, 22 (3) (1998) 393–406.

R. Staib, 'After Time Cost Quality and Scope, Now it is Time to Manage the Earth', *Australian Project Manager*, 19 (4) (1999) 13.

R. Staib, 'Archaeological Assessment in the Rouse Hill Urban Release Area', *Australian Archaeology*, 54 (2002) 22–28.

R. Staib, 'Environmentally sustainable design management, urban water infrastructure', *Australian Journal of Multi-disciplinary engineering, Engineers Australia*, 1 (1) (2003) 9–15.

R. Staib, 'Water Infrastructure & Archaeology in the Rouse Hill Urban Release Area', *Australian Planner*, 40 (1) (2003a) 21–26.

Standards Australia, *Risk Management, Australian/New Zealand Standard AS/NZS 4360: 2004*, (Sydney: Standards Australia, 2004), p. 30.

M. Starik, 'Should Trees Have Managerial Standing? Toward Stakeholder Status for Non-Human Nature', *Journal of Business Ethics*, 14 (3) (1995a) 207–217.

M. Starik, 'Research on organisations and the natural environment', *Research in Corporate Social Performance and Policy*, (1995b) 1–41.

M. Starik and G. Rands, 'Weaving an integrated web: Multilevel and multisystem perspectives of ecologically sustainable organisations', *Academy of Management Review*, 20 (4) (1995c) 908–935.

S. Sterling, *Sustainable Education: Re-Visioning Learning and Change*, (Foxhole: Green Books, 2001), p. 94.

H. Stretton, *Economics: a new introduction*, (Sydney, Australia: UNSW Press, 1999), p. 852.

R. Sullivan, 'Assessing the Acceptability of Environmental Risk – A Public Policy Perspective', *Australian Journal of Environmental Management*, 5 (1998) 62–70.

R. Sullivan, 'If Quantified Risk Assessment is the Answer, Are We Asking the Wrong Question?' *The APPEA Journal*, 40 (1) (2000) 635–642.

R. Sullivan and A. Hunt, 'Risk Assessment: The Myth of Scientific Objectivity', *Environmental and Planning Law Journal*, 16 (6) (1999) 522–530.

R. Sullivan and H. Wyndham, *Effective Environmental Management, Principles and Case Studies*, (Crows Nest, Sydney: Allen & Unwin, 2001), p. 246.

S. Sweet, N. Roome and P. Sweet, 'Corporate environmental management and sustainable enterprise: the influence of information processing and decision styles', *Business Strategy and the Environment*, 12 (4) (2003) 265–277.

Swiss Re, *Environmental Insurance for Enterprises*, (Zurich: Swiss Re, 1999), http://www.swissre.com/, viewed 11 May 2004.

Swiss Re, *The Insurability of Ecological Damage*, (Zurich: Swiss Re, 2003), www.swissre.com, viewed 11 May 2004.

Sydney Morning Herald, *Index to Sydney Morning Herald*, (Sydney, Australia: John Fairfax and Sons, 1964 to 1991), p. various.

Sydney Morning Herald, *Index to Sydney Morning Herald, 1987 to 2002 on Factiva*, (Sydney, Australia: John Fairfax and Sons, 2004), viewed 11 April 2004.

Sydney Water Corporation, *Towards Sustainability*, (Sydney: Sydney Water, 2002), http://www.sydneywater.com.au/html/environment/tsr/footprint.html, viewed 8 May 2004.

R. Taplin, 'Greenhouse Policy Development and the Influence of the Climate Change Prediction Timetable: An Australian Perspective', in T. S. Driver and G. P. Chapman, *Time-Scales and Environmental Change*, (London: Routledge, 1996), p. 275.

R. Taplin, 'Australian Experience with "New" Environmental Policy Instruments', *Annual Convention of the German Political Science Association on Environmental Policy and Global Change: Governance for Industrial Transformation 5–6 December*, (Berlin: Free University of Berlin, 2003).

P. Taylor, 'The ethics of respect for nature', *Environmental Ethics*, 3 (1981) 197–218.

Telstra, *Telstra's Consultative Council*, (Sydney: Telstra, 2004), www.telstra.com.au/tccc/index.htm, viewed 22 September 2004.

K. Tews, P. Busch and H. Jörgens, *The Diffusion of New Environmental Policy Instruments, Workshop No.1: New Environmental Policy Instruments, ECPR Joint Sessions of Workshops*, (Grenoble: ECPR, 2001), http://www.essex.ac.uk/ECPR/events/jointsessions/paperarchive/grenoble/ws1/tews_etal.pdf, viewed 26 December 2004.

The World Bank, *Pollution Prevention and Abatement Handbook 1998: Toward Cleaner Production*, (Washington: The World Bank, 1999), p. 472.

I. G. Thomas, *Environmental impact assessment in Australia: theory and practice 2nd ed.*, (Leichhardt, N. S. W: Federation Press, 1998), p. 258.

I. G. Thomas, *Environmental impact assessment in Australia: theory and practice*, (Leichhardt, N. S. W: Federation Press, 2001), p. 330.

I. G. Thomas and M. Elliott, *Environmental impact assessment in Australia: theory and practice 4th ed.*, (Leichhardt, N. S. W: Federation Press, 2005), p. 344.

K. Thomas, *Man and the Natural World: a History of the Modern Sensibility*, (New York: Pantheon Books, 1983), p. 425.

D. Throsby, 'Conceptualising Heritage as Cultural Capital', *Heritage Economics Conference: Challenges for heritage conservation and sustainable development in the 21st century*, (Canberra: Australian Heritage Commission, 2000) 10–17.

S. Thurwachter, *Environmental Value Analysis: Evaluating manufacturing product and process design trade-offs*, PhD, (Berkeley: University of California, 2000), p. 204.

R. S. Tibben-Lembke, 'The Impact of Reverse Logistics on the Total Cost of Ownership', *Journal of Marketing Theory & Practice*, 6 (4) (1998) 51–60.

T. Tietenberg, *Environmental and Natural Resource Economics (3rd ed)*, (New York, USA: HarperCollins Publishers Ltd, 1992), p. 678.

T. Tietenberg, *Environmental Economics and Policy*, (Meading, Mass.: Addison-Wesley, 1998), p. 460.

C. Tisdell, *Ecological and Environmental Economics: selected issues and policy responses*, (Cheltenham, UK: Edward Elgar Publishing Ltd, 2003), p. 361.

J. A. Todd and M. A. Curran (eds), *Streamlined Life Cycle Assessment: A final report from the SETAC North America Streamlined LCA Workgroup*, (Pensacola, Florida: Society of Environmental Toxicology and Chemistry, 1999), p. 31.

P. A. Tom, 'From dirt to dollars', *Waste Age*, 30 (8) (1999) 54–62.

Toyota, *Toyota Hybrid System*, (Japan: Toyota Motor Corporation, 2004), http://www.toyota.co.jp/en/tech/environment/ths2/index.html, viewed 17 December 2004.

H. Tsoukas, 'David and Goliath in the risk society', *Organisation*, 6 (1999) 499–528.

R. K. Turner, D. Pearce and I. Bateman, *Environmental economics: an elementary introduction*, (Hemel Hempstead, UK: Prentice Hall/Harvester Wheatsheaf Publishing, 1994), p. 328.

UNESCO, *Education for Sustainability, From Rio to Johannesburg: Lessons learnt from a decade of commitment*, (Paris: United Nations Educational Scientific and Cultural Organization, 2002), p. 46.

UNESCO, *Water for People – Water for Life – The United Nations World Water Development Report*, (Paris: UNESCO, 2003), http://www.unesco.org/water/wwap/wwdr/index.shtml, viewed 16 July 2004.

UNFCCC, *Caring for Climate, a guide to the climate change convention and the Kyoto Protocol*, (Bonn, Germany: UNFCCC Secretariat, 2003), http://unfccc.int/resource/cfc_guide.pdf, viewed 27 June 2004.

United Nations (UN), *Universal Declaration of Human Rights*, (Geneva: United Nations, 1948), http://www.un.org/Overview/rights.html, viewed 15 November 2004.

United Nations (UN), *The Rio Declaration on Environment and Development*, (Nairobi: United Nations, 1992), http://www.unep.org/Documents/Default.asp?DocumentID=78&ArticleID=1163, viewed 15 November 2004.

United Nations (UN), *Cleaner Production, a Training Resource Package*, (Paris: United Nations Environment Program, 1996), http://www.uneptie.org/pc/cp/library/catalogue/cp_training.htm, viewed 29 July 2004.

United Nations (UN), *Global Challenge, Global Opportunity, Trends in Sustainable Development*, (Geneva: United Nations, 2002a), www.johannesburgsummit.org, viewed 14 July 2004.

United Nations (UN), *World Population Prospects: The 2002 Revision and World Urbanization Prospects, Population Division of the Department of Economic and Social Affairs of the United Nations Secretariat*, (Geneva: United Nations, 2002b), http://esa.un.org/unpp, viewed 14 July 2004.

United Nations (UN), *Integrated Environmental and Economic Accounting 2003 (SEEA 2003)*, (New York: United Nations Statistical Division, 2003), http://unstats.un.org/unsd/envAccounting/seea.htm, viewed 16 June 2004.

United Nations (UN), *Treaties and Ratification*, (Nairobi, Kenya: United Nations Environment Program, 2004a), http://www.unep.org/ozone/Treaties_and_Ratification/index.asp, viewed 27 June 2004.

United Nations (UN), *Cleaner Production Activities Internet Site*, (Paris, France: United Nations, 2004b), http://www.uneptie.org/pc/cp/home.htm, viewed 27 July 2004.

United Nations Commission on Human Rights (UNCHR), *Economic, Social and Cultural Rights, Annexure: Draft Norms on Responsibilities of Transnational Corporations & Other Business Enterprises with Regard to Human Rights*, (Geneva: United Nations, 2002), http://www.unep.org/Documents/Default.asp?DocumentID=78&ArticleID=1163, viewed 15 November 2004.

United Nations Environment Program (UNEP), *World Summit Sustainable Development – Agenda 21*, (Johannesberg, South Africa: United Nations, 2002), http:// www.unep.org/, viewed 12 December 2004.

United States Environmental Protection Agency, *An Introduction to Environmental Accounting as a Business Management Tool: Key Concepts and Terms reproduced by The Chartered Association of Certified Accountants*, (Washington D.C.: EPA, 1995), p. 39.

USA Environmental Protection Agency (USA EPA) and Science Applications International Corporation, *Introduction to LCA (LCAccess–LCA 101. 2001)*, 2001, http://www.epa.gov/ ORD/NRMRL/laccess/LCA101 printable.pdf, viewed 12 December 2004.

USA EPA, *Guidelines for Preparing Economic Analyses*, (Washington, D.C.: US Environmental Protection Agency, 2000), p. 227.

R. van Berkel, 'Cleaner Production in Australia: Revolutionary Strategy of Incremental Tool?', *Australian Journal of Environmental Management*, 7 (3) (2000) 132–146.

R. van Berkel, E. Willems and M. Marije Lafleur, 'Development of an industrial ecology toolbox for the introduction of industrial ecology in enterprises', *Journal of Cleaner Production*, 5 (1–2) (1997) 11–25.

J. van den Bergh, *Ecological, Economic and Sustainable Development: theory, methods and applications*, (Cheltenham: Edward Elgar, 1996), p. 328.

J. Van Mierlo, L. Vereecken, G. Maggetto, V. Favrel, S. Meyer and W. Hecq, 'Comparison of the environmental damage caused by vehicles with different alternative fuels and drivetrains in a Brussels context', *Proceedings of the Institution of Mechanical Engineers – Part D – Journal of Automobile Engineering*, 217 (7) (2003) 583–594.

D. Vaughan, P. Scott and C. Mickle, 'Environment: What do Europe's Boardrooms think?', *Greener-Management International*, 7 (1994) 28–35.

C. Vehar, *The North American Database: Furthering Sustainable Development By Improving Life Cycle Assessment*, (Washington DC.: WISE Journal of Engineering and Public Policy (5), 2001), http://www.wise-intern.org/journal01/christinevehar2001.pdf, viewed 12 December 2004.

I. Vickers, 'Cleaner production: organizational learning or business as usual? An example from the domestic appliance industry', *Business Strategy and the Environment*, 9 (4) (2000) 255–268.

J. Viljoen and S. Dann, *Strategic Management, 4th Edition*, (Frenchs Forest, NSW: Pearson Education Australia, 2003), p. 468.

M. Wackernagel, L. Onisto, A. C. Linares, I. S. L. Falfán, J. M. García, A. I. S. Guerrero and G. M. G. S., *Ecological Footprints of Nations, How Much Nature Do They Use? – How Much Nature Do They Have?*, (Canada: The Earth Council Centre for Sustainability Studies, circa 1997), http://www.ecouncil.ac.cr/, viewed 22 May 2004.

K. Walker, *The Political Economy of Environmental Policy: An Australian Introduction*, (Sydney: University of New South Wales Press, 1994), p. 349.

A. Weale, *The New Politics of Pollution*, (Manchester: Manchester University Press, 1992), p. 227.

R. H. Weigel and L. S. Newman, 'Increasing Attitude Behaviour correspondence by Broadening the scope of the Behavioural Measure', *Journal of Personality and Social Psychology*, 33 (1976) 793–802.

E. Weizsacker, A. Lovins and L. Lovins, *Factor Four: Doubling Wealth, Halving Resource Use*, (London: Earthscan Publications, 1998), p. 358.

R. Welford, *Corporate Environmental Management 2, Culture and Organisations*, (London: Earthscan Publications Ltd, 1997), p. 192.

R. Welford, *Corporate Environmental Management 1, Systems and Strategies*, (London: Earthscan Publications Ltd, 1998), p. 270.

R. Welford, *Corporate Environmental Management 3, Towards Sustainable Development*, (London: Earthscan Publications Ltd, 2000), p. 184.

L. White, 'The Historical Roots of our Ecologic Crisis', *Science*, 155 (3767) (1967) 1203–1207.

A. Wilkinson, M. Hill and P. Gollan, 'The sustainability debate', *International Journal of Operations and Production Management*, 21 (2001) 1492–1502.

M. Williams, 'Why doesn't the government respond to the participating public?' *Participation Quarterly (International Association for Public Participation)*, February (2004) 5–8.

I. Wills, *Economics and the Environment*, (St Leonards, Australia: Allen and Unwin, 1997), p. 340.

C. Wood, *Environmental Impact Assessment: a comparative review*, (Harlow, London: Longman Scientific & Technical, 1995), p. 337.

J. L. Woodka, 'Sentencing: Personal Liability of Corporate Executives for Environmental Crimes', *Tulane Environmental Law Journal*, 5 (1992) 635–662.

J. Woosley, *EMS Development Course for Government Agencies*, (Franklington, NC: North Carolina Department of Environment and Natural Resources, 2001), http://www.p2pays.org/iso/public/govcoursesched.asp#preclass, viewed 12 December 2004.

World Business Council for Sustainable Development, *About the WBCSD*, (Geneva: WBCSD, 2004), http://www.wbcsd.ch, viewed 22 May 2004.

World Commission on Environment and Development (WCED), *Our Common Future*, (Oxford: Oxford University Press, 1987), p. 400.

World Resources Institute, *Compact Disc of World Development Indicators including: population, water, food, energy*, (Washington, D.C.: International Bank for Reconstruction and Development/World Bank, 2003).

H. J. Wu and S. C. Dunn, 'Environmentally responsible logistics systems', *International Journal of Physical Distribution & Logistics Management*, 25 (2) (1995) 20.

J. M. Yarwood and P. D. Eagan, *Design for Environment Toolkit: A Competitive Edge for the Future*, (Minnesota, USA: Minnesota Office of Environmental Assistance, 1998), http://www.moea.state.mn.us/berc/DFEtoolkit.cfm, viewed 16 December 2004.

D. Yencken, 'Governance for Sustainability', *Australian Journal of Public Administration*, 61 (2) (2002) 78–90.

K. Yongvanich and J. Guthrie, 'Extended Performance Reporting Framework: A Form Of Sustainability Reporting', *The Inter-Disciplinary CSR Research Conference, 22–23 October 2004*, (University of Nottingham, Nottingham, UK: University of Nottingham, 2004).

J. Young, *Post-Environmentalism*, (London: Belhaven 1, 1992), p. 225.

O. Young, *Global Governance: Drawing Insights from the Environmental Experience*, (Cambridge: MIT Press, 1997), p. 364.

P. G. Zimbardo and M. R. Lieppe, *The Psychology of Attitude Change and Social Influence*, (Philadelphia, USA: Temple University Press, 1991), p. various paging.

Index

Note: Pages in bold are the principal references.

Printed in the United States
By Bookmasters